ANTIAESTHETICS

SYNTHESE LIBRARY

STUDIES IN EPISTEMOLOGY,

LOGIC, METHODOLOGY, AND PHILOSOPHY OF SCIENCE

VOLUME 174

PAUL ZIFF
The University of North Carolina at Chapel Hill

ANTIAESTHETICS

An Appreciation of the Cow with the Subtile Nose

D. REIDEL PUBLISHING COMPANY
A MEMBER OF THE KLUWER ACADEMIC PUBLISHERS GROUP
DORDRECHT / BOSTON / LANCASTER

Library of Congress Cataloging in Publication Data

Ziff, Paul, 1920-
Antiaesthetics, an appreciation of The Cow with the subtile nose

(Synthese library; v. 174)
Bibliography: p.
Includes index.
1. Aesthetics–Addresses, essays, lectures. I. Title.
BH39.Z54 1984 111'.85 84–17854
ISBN 90–277–1773–7

Published by D. Reidel Publishing Company,
P.O. Box 17, 3300 AA Dordrecht, Holland.

Sold and distributed in the U.S.A. and Canada
by Kluwer Academic Publishers,
190 Old Derby Street, Hingham, MA 02043, U.S.A.

In all other countries, sold and distributed
by Kluwer Academic Publishers Group,
P.O. Box 322, 3300 AH Dordrecht, Holland.

Printed in The Netherlands

For Matthew

TABLE OF CONTENTS

PREFACE

Although various sections of this work have been published separately in various journals and volumes their separate publication is wholly attributable to the exigencies of life in academia: the work was devised as and is supposed to constitute something of an organic unity.

Part II of 'The Cow with the Subtile Nose' was published under the title 'A Creative Use of Language' in *New Literary History* (Autumn, 1972), pp. 108-18. 'The Cow on the Roof' appeared in *The Journal of Philosophy* LXX, No. 19 (November 8, 1973), pp. 713-23. 'A Fine Forehand' appeared in the *Journal of the Philosophy of Sport,* Vol. 1 (September, 1974), pp. 92-109. 'Quote: Judgements from Our Brain' appeared in *Perspectives on the Philosophy of Wittgenstein,* ed. by I. Block (Oxford: Basil Blackwell, 1981), pp. 201-211. 'Art and Sociobiology' appeared in *Mind* (1981), Vol. XC, pp. 505-520. 'Anything Viewed'appeared in *Essays in Honour of Jaakko Hintikka,* ed. by Esa Saarinen, Risto Hilpinen, Illkka Niiniluoto and Merrill Provence Hintikka (Dordrecht, Holland and Boston, Massachusetts: D. Reidel Publishing Co., 1979), pp. 285-293. 'How I See Philosophy' appeared in *The Owl of Minerva,* ed. by C. J. Bontempo and S. Jack Odell (New York: McGraw-Hill Book Co., 1975), pp. 223-5. All the remaining parts are also forthcoming in various journals and volumes. I am grateful to Bradley E. Wilson for the preparation of the index.

Chapel Hill

PAUL ZIFF

George, you old debutante,
How did you get in the Army?

I

THE COW WITH THE SUBTILE NOSE

I

Since for me the notion of intention is senseless or as senseful as such kindred concepts as spirit and ghost I cannot say what the essay following these few words is intended to say. What is said there is perhaps plain enough. I have however elected to underline as it were certain of its aspects.

Being an entity this essay has of course diverse aspects. One such is its obvious division into three parts chapters so called. This division is from a philosophic point of view gratuitous. A cautious concession to anyone's flagging attention failing eyes nagging thirst and familiar conceptual scheme. I would not suggest that any cut is philosophically arbitrary. The space the gap dividing 'the' from 'cow' in the phrase 'the cow' has its *raison d'être* in the sense of what is being said which would not be what it is if that gap were not where it is and what it is resolutely planted arms akimbo between. But the difference between speaking of "the cow with a subtile nose" and "thecow with a subtile nose" need not detain us. I for one freely admit that I have no idea what a thecow is and certainly I have never seen one if it is something to be seen which of course is far from clear and need not be granted by anyone.

That some organization is wanted is not to be denied but what kind and how much is another matter distinctly a horse of another color. One man's organization may well be another man's meat. For example one way to pack a large box with small boxes is carefully to place the small boxes one by one flat alongside one another and so forth in the large box. Another way is to throw the small ones higgledy piggdledy or higgeldy piggeldy or higgeldy piggledy or higgledy piggeldy or even higgdledy piggdledy or higgdeldy piggdledy or higgdledy piggdeldy or higgdeldy piggdeldy and so forth into the large box. The results will not be the same. Anyone can see that. But in each case the small ones will be in the large one though not necessarily the same number of small ones or even the same small ones. Anyway an essay in aesthetics or even in antiaesthetics is not a large box packed with small goodies or a small box packed with large goodies or any kind of box at all packed with anything at all. (There's really no need to worry about boxes here. We'll get to them soon enough. Meanwhile we can persist as cheerful introspective organic automata aesthetically oriented God bless us every one!)

There is a clear cut effective procedure for forcefeeding a goose. All one needs is a consenting goose and a funnel and feed and a ramming device. What is the algorithm for understanding art? Besides being a civilized being not given to gluttony you may not covet a goose with a disastrously enlarged liver enlarged beyond all understanding and perhaps before entropy grabs us one and all you and me and that goose you may prefer to consort with that goose to consider its ways and for that there are as yet no procedures available.

2

What is a creative use of language? A fine thing perhaps for to say of anyone that his is a creative use is presumably to praise but what is it?

A use of words any ordinary use of words? Is that possible? One expects a creative use to be somehow special. Or is 'a creative use' like 'a four legged aoudad'? Any ordinary aoudad is a four legged aoudad an everyday aoudad not some special kind. Wordsworth wrote and possibly took it upon himself to say right out loud

> Who weeps for strangers? Mary wept
> For George and Sarah Green.

An ordinary use of words. Yes. Wordsworth was on occasion a poet but that does not matter. His use of words here was ordinary. Was he in saying what he said being creative?

He made sounds or if he managed to avoid the utterance then he made marks on paper. If making were here creating then here Wordsworth was being a creator. And making sometimes is and when it is creating is easy. It calls for no gift. Bulllike roars in a hushed auditorium can make and create a disturbance. An inadvertently upset soup can make and create a mess. A medical person may have made and created an available moment in which to examine a thorax. A thorax possessor may have made and created free time in which to have it examined. These acts of creating need not detain one in a study of creativity. That someone is always creating a mess does not incline one to conclude that he is a creative type. Not even if he does it with ink on paper or paint on canvas.

When Wordsworth said "Who weeps for strangers? Mary wept / For George and Sarah Green." he made sounds. Did he in making such sounds create sounds? He could have. Given facilitating conditions he could have created echoes. What if there had been no echoes? Waves breaking on a shore make sounds. Do they in that perform a creative act? Perhaps people only or anyway only organisms perform acts. Then what is in question is not an act of the waves for there is none. The action of the waves is that creative? Windblown grinding sands have created fantastic shrieks in Monument Valley. The action of the winds may be creative. It does not follow that making sounds is creating but sometimes it is. The action of the breeze in making the leaves of a forest rustle makes and creates a rustling sound. The wash and slide of the waves on a shore make and create layer upon layer of sound.

Is making sounds creating? It need not be. If a man sits in a corner mumbling yes an ordinary use of words he may make sounds and create nothing. This is hard to understand and it is something to think about and things can seem uncertain but they won't if one thinks about it. Why can't he do what the wind can do? A man sits in a corner mumbling. Then he's making noise and creating noise. No. Noise is unwanted sound and maybe there's no one around to unwant the sound. Then at least he's making sounds and so he's creating sounds. No. The wind through the leaves creates rustling sounds. He in the corner makes sounds and perhaps creates nothing. Why can't he?

Sometimes one doesn't just make sounds but creates sounds and what is the difference? Of course it depends on the sounds. That is an important difference between sounds and a racket. If one makes a racket then it doesn't matter what any further character of the racket is for in making that racket one has created something if only a racket. Whereas if one makes sounds it does matter what the further character of the sounds is and only if the sounds have a special character will one in making the sounds have created something.

If the mumbling man is creating sounds then one wants the sounds he is making to be somehow special sounds. Adolphe Sax created very special sounds with his saxophone but the mumbling man in the corner is making sounds and creating nothing. The sounds he makes any or just about any can make while the sounds the wind makes only the wind or just about only the wind can make or anyway that is the way it seems. And then there are millions of mumbling men but the winds are not so numerous. Mistral Sirocco Chinook. One could hope to name them all.

More. The sounds the man makes are sounds that humans make. Thinking of the sounds humans make as constituting a set they don't if one takes 'set' literally but don't take it literally this set is sortable into many varied and distinct proper subsets. There are all sorts of human sounds and we humans are connoisseurs of human sounds. We are not such connoisseurs of sounds the winds make possibly because they do not matter as much to us.

A merely mumbling man makes sounds but if the sounds he makes are not somehow special then even though he makes sounds he does not in making sounds create sounds. Certainly the same is true of Wordsworth when he said "Who weeps for strangers? Mary wept / For George and Sarah Green." In saying that he made sounds but he need not have created sounds. Did he create anything?

When Wordsworth said what he did about Mary and George and Sarah Green he uttered an eleven word sentence. Maybe the sound sequence constituted by the utterance of his eleven word sentence had never been produced before. Doesn't that mean that the sounds he made were special sounds in that they were absolutely novel and so in making these sounds he was creating an absolutely novel sound sequence? No.

If that is a case of making something special then there is nothing special about something special. And if those lines testify to the creativity of their author then there isn't much to being a creative writer and it is easy to be one. Just spell out a new sentence grammatical or otherwise and call it a poem. There are more composable word sequences than anyone needs. With a generous vocabulary of 400 000 words and allowing repetitions and ignoring grammatical restraints as poets sometimes do one can come up with 400 000 to the eleventh power number of eleven word sentences. If one were to voice these sentences at the rate of one million per second it would take in excess of 10 to the 40th power number of years to voice them all. So if we so babbled on all of our time we could all of us be creative all of our time. Not only on this account would virtually all of us prove to be creative types but extremely simple suitably programmed special purpose digital computers could then be classed creative entities. For though no computer for want of world enough and time could print out all of the 400 000 to the 11th power number of eleven word sentences even a modest computer could make a beginning and almost at once come up with novel sentences exemplifying novel sound sequences.

Some seem unduly taken with novelty. N. Chomsky the linguist says that

> The most striking aspect of linguistic competence is what we may call the 'creativity of language', that is, speaker's ability to produce new sentences, sentences that are immediately understood by other speakers although they bear no physical resemblance to sentences which are 'familiar' .... In fact, even to speak of the hearer's 'familiarity with sentences' is an absurdity. Normal use of language involves the production and interpretation of sentences that are similar to sentences that have been heard before only in that they are generated by the rules of the same grammar, and thus the only sentences that can in any serious sense be called 'familiar' are clichés or fixed formulas of one sort or another.[1]

That speakers can and do produce new sentences is true. That these new sentences "bear no physical resemblance" to sentences which are familiar is not true. When Browning said

> Irks care the crop-full bird? Frets doubt the mawcrammed beast?

he brought forth a mouth filling sequence of sounds instancing a novel sound pattern. But it bears a distinct resemblance to the more ingratiating

> Cock a doodle do! Cock a doodle do!

as well as to

> Peter Piper picked a peck of pickled pepper.

Neither is it true that the normal use of language involves the production of sentences that are similar to sentences that have been heard before "only in that they are generated by the rules of the same grammar". The normal use of language involves the production of sentences that are strikingly similar to sentences that have been heard before. A novel sentence is likely to be similar to one that has been heard before in any or all of various respects. Phonology morphology syntax semantics may each and all establish an aspect of similarity. The wholly novel sentence that bears no linguistic similarity to any sentence heard before is an impossibility. It could not exist.

Mere novelty is not significant. Novel sentences are readily composable by standard data processing means. Absolute novelty is hardly hard to come by. Physiologists seem to assure us that each person is physiologically unique. Botanists tell us of each blade of grass's individuality though cosmologists do sometimes mutter uncomfortably about our universe being caught perhaps in time's rut endlessly repeating itself. Absolute novelty seems so utterly unimportant uninteresting.

> Henry sats in de bar & was odd,[2]

A part time hermit poet H. Furry out in Yucca Flats wrote the identical line and was amazed to find it in Berryman's *Dream*

Songs. He wrote to the *Tucson Times* pointing out the coincidence. He had never met or read Berryman. Berryman had never seen the hermit's line so an unlikely event occurred. Of course the rest of Furry's production was nothing at all like the 5th *Dream Song*. Is that interesting? No that isn't interesting when what's in question is creativity. Absolute novelty is of no consequence which isn't to deny that some sort of novelty may be significant.

Novelty of some sort is wanted. If one makes plans to play tennis tomorrow as he does every day then there need be nothing novel about his plans. In making such he perhaps creates nothing at all. Whereas if one makes plans to corner the grain market by means of a novel manoeuvre then he has created a novel plan to corner the grain market.

But mere novelty is not enough. What if one plans to play tennis on Friday the 28th of April? If each day one plans to play tennis tomorrow then one's plan to play tennis tomorrow is not novel. But if tomorrow is Friday and one plans to play tennis tomorrow then possibly one plans to play tennis on Friday and that is not something that one does everyday. And if Friday is the 28th of April and one plans then to play tennis then quite possibly that is absolutely unique. Even so one need not have created anything. Not every aspect of a plan is apt to concern us. To plan to play tennis tomorrow is not exactly the same as to plan to play tennis Friday even if tomorrow is Friday and is not exactly the same as to plan to play tennis on Friday the 28th of April even if tomorrow is Friday the 28th of April. But these factors are not apt to matter overmuch. If one does plan to play tennis tomorrow and if tomorrow is Friday the 28th of April does that signify? Will that lead one to say 'I thought the plan was everydayish commonplace but now I see that it is extraordinary brilliant. That tomorrow is Friday the 28th of April makes all the difference'? Then perhaps in making the plan one created something. That is not the way it is in the ordinary run of things.

Irrelevant novelty is a common occurrence in the everyday use of language. "For there, in a gesture of economy and perhaps utility, Los Angeles County has installed 900 plastic peperomias, pittosporums and ti palms, among other flora, to decorate the roadway".[3] Possibly prior to its having been written by the person of *Time* magazine this sentence had never been written before. Then hasn't the person in writing this created a novel sound pattern? No. He has produced a sentence which does exemplify a certain sequence of sounds. It does not follow that he has created a novel sound pattern or any sound pattern at all.

A man does what he does but that does not mean that whatever is then done is what he did. A man threw a rope down on a square floor divided into four equal quadrants. The curve of the rope then constituted the graph for some function. It was bound to. Call that function '*phi*'. On being asked 'How did you ever manage to throw a rope down in such a way that it would constitute the graph of *phi*?' he replied 'There was nothing to it. I just threw it.' 'Then do it again!' No way! He threw the rope down in such a way that it happened to graph *phi* but not in such a way that it would graph *phi* even though it did.

The *Time* magazine writer produced a sentence which happened to exemplify a certain sequence of sounds. It was bound to exemplify some sequence or other. He happened to produce a certain sequence of sounds. Happening to produce something is not the same as creating anything. That the writer simply happened to produce the sequence of sounds in question can be seen in the fact that his use of words was such that he would be deemed by one and all and himself to have said the same thing if instead of writing "900 plastic peperomias, pittosporums and ti palms" he had written '900 plastic pittosporums, peperomias and ti palms'. Such a deviation from his original alphabetic ordering of the floral terminology would be of no significance.[4]

Sometimes when some speak they create distinct sound patterns. That phrase 'sound patterns' does not perhaps do justice to what I intend. So I shall speak of "melodic sound patterns" though that phrase has its misleading aspects. I am not concerned to evoke shades of "Ulalume". Milton created melodic sound patterns when he first said

> Yet once more, O ye laurels, and once more
> Ye myrtles brown, with ivy never sere,
> I come to pluck your berries harsh and crude,
> And with forced fingers rude
> Shatter your leaves before the mellowing year.

And Tennyson created melodic sound patterns when he wrote said sang

> The splendor falls on castle walls
> And snowy summits old in story;
> The long light shakes across the lakes,
> And the wild cataract leaps in glory.
> Blow, bugle, blow, set the wild echoes flying,
> Blow, bugle; answer, echoes, dying, dying, dying.

Using words to create melodic sound patterns is what poets do when their words are close to music. It is not what Wordsworth did when he said what he did about Mary and George and Sarah Green. It is not what the *Time* magazine writer did in producing a sentence so readily altered without significant effect. In contrast consider the havoc that would be wrought were one to reorder Milton's floral terms. In place of

> Yet once more, O ye laurels, and once more
> Ye myrtles brown, with ivy never sere,

one would have

Yet once more, O ye myrtles brown, with ivy
Never sere, and once more ye laurels,

Ignoring the unsubtle prosodic change to begin with the alliterative pair 'more' 'myrtles' would get *Lycidas* off to a dead start. Milton's melody would have been flawed.

Doesn't tampering with word order in any case destroy the precise melodic sound pattern created? That hardly seems plausible when one ponders the timely example just given and Wordsworth's rural ditty was surely mute. But there are other cases. Julia Moore the Sweet Singer of Michigan supplies us.

While eating dinner, this dear little child
  Was choked on a piece of beef.
Doctors came, tried their skill awhile,
  But none could give relief . . . .
Her friends and schoolmates will not forget
  Little Libbie that is no more;
She is waiting on the shining step,
  To welcome home friends once more.

In writing Little Libbie's epitaph Julia Moore may have been the first to exemplify such a sequence of sounds. Yet the plain and hardly simple fact of the matter is that not just any sequence of sounds can sensibly be characterized as constituting a melodic sound pattern. Just so exactly so not just any sequence of tones can sensibly be characterized as constituting a melody.

What is a melody? That's hard to say. But this much seems certain. Melodies are constituted by only some but not by all finite sets of successive musical tones. Melody is essentially the horizontal element of musical texture in contrast with harmony the vertical element. Musical tones are characterized by pitch and duration. A specific melody can be thought of as an ordered finite set of ordered pairs. Strictly speaking this characterization of a melody as an ordered finite set of ordered pairs is a drastic oversimplification. A melody may be transposed adorned with grace notes and so forth and still remain the same melody. Thus a melody is not a specific ordered set but at best an indeterminate collection of ordered sets of ordered pairs. But these complications do not matter to the point that I am concerned to make.

Let us restrict our attention to ordered sets of 30 pitches of equal duration that can be sounded on a modern piano of 88 keys. The total number of such sets is 88 to the 30th power. Of this set of sequences only a proper subset can sensibly be supposed to constitute a set of melodies. That sequence constituted by the pair of pitches a^4 and b^4 repeated 15 times with each note of equal duration can hardly sensibly be supposed to be a melody. But which members of this enormous set of sequences are melodies and which are not?

There is no clear answer to this kind of question. Of some sequences one can readily say that yes they are melodies. Of others no they are not. Of others they sort of are or they sort of are not or one simply can't say. So as one is often compelled to conclude there will be a spectrum of cases. To make matters worse or anyway worse for those who have a taste for computation there is not at the present state of the arts any way of computing whether or not a given sequence is a melody. To try it on a well tuned ear is all one can do. And this indicates that whatever else computers inorganic automata mechanisms can do for the time being they cannot be creative composers. Because though they might happen to print out a melody it could not be a case of creating a melody because they could not know what they were doing in doing it because for the time being only a well tuned ear can recognize a melody. Only people have such ears.[5]

Why can't a computer compute whether or not a given sequence of tones constitutes a melody? I don't know the answer to this question but it is not a matter about which one must feel altogether blank. I mean that if one says as I say that this is something that at present no computer can do and maybe none will ever be able to that isn't to give up being sensible hardheaded scientific. It isn't to invoke ghosts souls and spiritual claptrap all in the name of art and creativity. Such an incapacity could be readily explicable. For example it could be the case that to do it either we or the computer would have to break what is conceivably an unbreakable code. To put the matter no doubt loosely and certainly figuratively suppose that to determine whether or not a given sequence of tones constitutes a melody we operate on the given sequence with a certain stored set of functions. I don't think we in fact do exactly this but possibly what we do do has a similar logical character and anyway suppose we did. Then if the operation on the given sequence yielded some designated value it could be classed a melody and if not not. A computer then could determine whether or not a given sequence was a melody only if it had access to the stored set of functions. But conceivably neither we could supply nor the computer could have such access and conceivably it never could have such access. For the physical realization of the stored functions could be inaccessible neutral structures corresponding to innate features of the human mind in virtue of which we class some but not all sequences of tones melodies.

> While eating dinner, this dear little child
> Was choked on a piece of beef.

The Sweet Singer of Michigan here failed to create a melodic sound pattern. One can't show by computation that this is so but it is so and anyone with a well tuned ear can hear its unmelodic bleat.

To create something it seems that one must characteristically anyway make something that meets the following vague conditions. What one makes must be relatively if not absolutely novel. It must be relevantly novel and not merely novel because of inessential features. It must be somehow something special and not special merely by being novel and not because of inessential features. Yes all this is vague and indefinite and certainly indecisive. And since 'creative' and 'create' seem to suffer from loose uses anyway even this vague characterization may fail to fit cases here and there.

But what it licenses does seem right. Suppose a man makes a crazy kind of coffee cup one filled with minute holes. In consequence as one drinks or tries to drink coffee from it the coffee slowly leaks out. Then one could claim that the man has created a new kind of coffee cup and one could deem it designed to inhibit excessive coffee drinking. Again consider a lazy inept tailor who made a new coat but made a mess of it a coat of crooked seams and sagging sides. The tailor might protest that far from being lazy and inept he was a truly creative type. He had created a coat for he had made a new coat that was something special. And if in consequence of his efforts crooked seams became the fashion and sagging sides all the rage then what seemed to be a lazy inept tailor could turn out to be a creator.

Is it then after all so easy to create? That depends on the field of one's endeavours. It is easy to create things of a kind that people have not been concerned to create. An ardent creator could create a typewriter that jams immediately to inhibit writing and automobiles with square wheels to prevent traffic jams and blank eye charts to cheer the blind. To be creative in the use of language is not all that easy. It is not hard to produce sentences and as has been indicated novelty is easy to come by. But it is harder and much harder to make new paragraphs or sentences or lines or phrases that have a genuinely special character and that are not simply trivially special because of inessential features.

To turn again to the Sweet Singer of Michigan when one reads her parting comment

And now, kind friends, what I have wrote,
  I hope you will pass o'er,
And not criticise as some have done
  Hitherto herebefore.

one realizes that one cannot deny that Julia Moore was a creative poet. Though she seems more prone to jingling than singing her lines are novel and have an altogether special character. Probably not the character she wanted them to have but they have a special character all their own. Again the Babu poet who wrote on the death of Queen Victoria

Dust to dust, and ashes to ashes,
Into the tomb the Great Queen dashes.

had an unmistakable creative flair. That the madness of the image probably escaped the poet is of no consequence. As D. B. Wyndham Lewis had remarked in his preface to *The Stuffed Owl* "good Bad Verse has an eerie supernal beauty."[6] In contrast however with the cases just considered when one ponders the lines

And when upon your dainty breast I lay
My wearied head, more soft than eiderdown.

by William Nathan Stedman one is not inclined to attribute any creative powers to the poet despite the unmistakable charm of the lines. Because most likely he didn't know what he had done and wouldn't have done it if he did.

A creative use of language calls for the production of novel sentences having some special character. One such special character is that of constituting a melodic sound pattern. Unfortunately no analysis of what constitutes a melodic sound pattern is at present available. But a creative use of language may display other kinds of special qualities. So one may look to Shakespeare's imagery Marvell's wit Donne's elaboration of metaphysical conceits. But again and unfortunately although many of us have a fairly clear and coherent conception of what fine imagery is and we can readily recognize wit and appreciate metaphysical conceits there are no available analyses of these notions. The questions 'Is that a fine image?' 'Is that wit?' 'Is that a coherent account?' do not today have computable answers which is not to say that we cannot in any way answer them. In some cases we can and in some we cannot and in some it is altogether unclear what we can or cannot do.

To complicate matters still further one may recognize that a work one is concerned with has an altogether special character without being able to provide any helpful description or specification of that character. This is a common enough occurrence in art criticism. To talk about what's special about a painting by Poussin is easy and one can talk on and on. To say what's special about a painting by Jean Dubuffet may be impossibly difficult. There often is nothing worth saying. 'The expression on the face of the person in the picture. Look at it!' 'What about it?' If one doesn't see it there's nothing to say to him. For that is what gives the work its special character. The same is true of some poems. Here is Berryman's *Dream Song 77*.

> Seedy Henry rose up shy in de world
> & shaved & swung his barbells, duded Henry up
> and p.a.'d poor thousands of persons on topics of grand
> moment to Henry, ah to those less & none.
> Wif a book of his in either hand
> he is stript down to move on.
>
> – Come away, Mr Bones.
>
> – Henry is tired of the winter,
> & haircuts, & a squeamish comfy ruin-prone proud national
> mind, & Spring (in the city so called).
> Henry likes Fall.

Hé would be prepared to líve in a world of Fáll.
for ever, impenitent Henry.
But the snows and summers grieve & dream;
thése fierce & airy occupations, and love,
raved away so many of Henry's years
it is a wonder that, with in each hand
one of his own mad books and all,
ancient fires for eyes, his head full
& his heart full, he's making ready to move on.

3

She walked into the museum lobby. She stopped. Standing still she looked for a long moment back at the revolving door going

whish

whish

whish

whish

The doorman had been polite. It was raining. He had held a large candy striped umbrella for her. He was patient even if disturbed. It had taken much time and he had been unsure where to position the umbrella. Not that the wind had been unpredictable. There was a lack of fit between her and his umbrella. He had shaped and reshaped his instrument to avoid such issues. No good. He never managed to anticipate future developments. If then only partial protection is to be afforded precisely where is it best applied? The doorman had put this problem to himself over and over again but again he could find no answer.

The passage had been accomplished. It had required her full prolonged attention. When there is much of one as there is of her revolving doors seem designed less for the prevention of drafts than for the establishment of some obscure equipoise principle of endozoic harmony. As much of one that enters this place that much of one must leave. She felt not well ill at ease the walls. She did not like the walls. They defined a finite space. Behind her were others and they will force a place for themselves at whose expense? Successors dwindle to predecessors and perhaps disappear. Limits everywhere. The image of herself as the eventual discrete excretion of a massive museum body was bewildering. Should she leave at once? But the thought of making another pass at that door was too much for her.

It was Sunday. Dubuffet was being shown in New York. The galleries were congested swollen brimful. Thousands of bodies seeped through partitions flooded corridors. They were spilling over guards and sculpture splashing off pictures soaking up name tags. She looked around with a vague distressed look. The pressure grew. To view paintings under such pressures. She felt impossible to appreciate. The conditions of perception in a public museum even any day are not conducive to an aesthetic experience. Our bars are better designed for what they are and what they would be. A warm stall a dim quiet place. It is perhaps no one's fault so perhaps there is no one to blame which doesn't make matters any better. Let's take this painting home to look at to wake to awhile. Can't. Can't afford it. His works sell for 40 50 150 thousands of dollars. Anyway it's not for sale. This is a public place belongs to the public. So nobody's gonna wake up with it and get familiar. Museums are alien guarded solemn places as comfortable as mausoleums more like churches or banks than whatever it is it would be nice they were like.

It would be nice if they had salons where a few works could be hosts would entertain over coffee or cocktails or whatever a small number not more than 20 guests would be invited to last 45 minutes by reservation only at scheduled hours designed and decorated for their hosts with seats and tables placed and fine carpets for guests wearing paper slippers. Floors would not be what they are and certainly not made of marble.

To walk on marble floors isn't easy. It isn't easy for anyone and certainly it's not easy for an ungulate. To backtrack somewhat it is incumbent on all of us to understand that anyway it is difficult being an ungulate. There is first of all the matter of hooves. Hooves aren't like soft feet but are of horn. This curved covering of horn affords some protection from the pebbles rocks stones that are to be found in pastures given over to cows. Grass and mud unlike museum floors are soft under hoof cushion and ease the shock each incautious step affords.

And then there is the weight. Eleven hundred pounds give or take a few hundreds. If we assume an even distribution of weight as she stands looking for she will stop to look at a book a picture a slide will walk and stop to look again and again if we assume that then something like two hundred seventy five pounds must be assigned to each leg. Since the edifice of the leg is erected on an unmistakably horny plinth it is horn which is to say hoof that bears the burden of her body.

There were some gasps. Would be viewers turned surprised faces. Glittering camera lenses flashed snapped. Some only a few gnashed many more clattered rolled astonished eyes and clicked teeth. There were also an arbitrary number 17 of titters 47 guffaws and any number of suppressed intakes and outputs of indiscriminate character.

The doorman had followed her in. He had been tending to her throughout the time it had taken to pass through the revolving door. The process had nurtured a proprietary air blowing about him in a cheerful breeze. He smiled in a confident fashion at those clicking their teeth. He asked politely if they would prefer castanets? Had they heard flamenco true flamenco in Barcelona pronounced *barthelona*? Nodding rapidly smiling a few times all in the best of humor he assured everyone present that if there was if he could be of any service he was there waiting to be wanted. Then looking uneasily in her direction uncertainly at the teeth clickers and in a peculiarly muffled uneasy thin scratchy voice he said that if he might mention it not that he wished to offend certainly not not that he was taking anything upon himself for after all it was a free country and people that want to click can click if that's what they want it's up to them. But with a sour look now she mind you couldn't really be expected to be cheered by any sort of sound that would put in mind a bull ring. This last phrase being brought forth in a small carefully wrapped worn with use but evidently once a brightly gaily colored paper parcel tied securely with bits of old blue string being carefully untied and undone taken from old tissue paper dust and dirt being carefully surely blown off before being fitted to his mouth and then brought out in barely a whisper so that probably no more than two or three of the closest to this mouth could have heard him. She certainly didn't or if she did she gave not the slightest indication.

Vets don't mind war stories. Why should she be bothered by rapping about bull fights? It's not as though she'd get it in the neck. That's for bulls. This was suggested by one of the would be viewers a long-haired dirty jeans whose pony tail needed combing. She oughta read Hemingway's *Death in the Afternoon* advised a short beard with cold eyes wired to his head with steel rims. The doorman's sour look seemed to relent melt a mite. An avid gleam of patience now appeared as one of his certain virtues. You don't understand he said looking around to see where she was. Assured by what he saw that where she was was out of earshot he visibly relaxed and loosened his jaw preparatory to freeing his mouth.

She had walked into the museum book store. Quietly. Not silently but at least quietly. Marble is not really what it is. Not marble but some composition that can be spread an artificial marble seeming substance deployed to meet public floor exigencies. It is nonetheless very hard. *Kalunkh kalunkh kalunkh*. Remarkably not a fantastic heavy voluminous sound not a crashing *THONK* not a fierce *clop clop*. She is light on her feet. That is surprising. Some ungulates are agile capable of deft swift movements. But not Bos. Capra of course. And anyone's darling Ammotragus lervia.[7] But Bos?

Cows are ruminants thoughtful meditative types. Occasionally given to gamboling but prone to pauses reflective interludes. They look then they think. They chew the matter over. They recollect. Dubuffet did see cows. She saw him seeing. He has said so.

On the outskirts of in 1954 Clermont-Ferrand. "I loved spending hours watching the cows and afterwards drawing them from memory, or even, but much more rarely, from life."[8] She recalls that he looked at cows. Just look at her and you can see at once that he looked at cows spent hours looking at cows. She looked at him.

He says "First of all I should say that the sight of this animal gives an inexhaustible sense of well-being because of the atmosphere of calm and serenity it seems to generate. I can also say that pastures and even merely the colour green – because of the cows, I suppose, by an unconscious association of ideas – has a comforting and soothing effect on me."[9]

Empathy you see said the doorman loudly in the lobby but she doesn't hear. Her long nose sways up and down sniffing a far away clover. Her eyes are turned inward to long ago pastures. There is this man standing staring. He is 53. Cows and green grass pastures. He is watching but cows do not move much. Munch or munch chew look silently. An occasional moo. Slow everything is slow the clouds quietly the leaves softly move. The light changes serene and calm. This old man stands staring. His feet do not quite reach the ground. Rough workmen's shoes cover his feet covered with coarse woolen socks rough denim blue pants sagging everywhere down to the shoetops. A gust of wind buffets him about but he holds to the railing fence sinewy hairy tendoned fingers clutching. The crows cry caw. The cows float in the meadow bouncing up and down in the light breeze.

There is this cow now standing staring. She has four legs four or more anyway udders one tail one head two eyes one nose long subtile two horns but where are her ears? She stands staring her tail hangs down but nothing hangdog there. She sees this old man hovering by the fence feet floating stirring shuffling in air. It is certainly an interesting sight. She has stopped chewing to look.

Empathy of course was what Papa was short on. He and they Hemingway he went on quoting Gertrude's (Stein) phrase with conspicuous approval rolling the phrase round and round his mouth enjoying his center ring juggling act he and they had no feeling for cows or for ungulates of any kind. Blasting around Afrika with an elephant gun blowing holes in ungulates in felines in mammals of every kind. No such son of a bitch solemnly stated the doorman with the plain stare and balanced stance of one about to deliver a henceforth to be received truth and rightly of course no such son of a bitch he repeated is in a position to appreciate Dubuffet's cow with the subtile nose.

Empathy is what is wanted and it can be cultivated. A cow is a person just like any other. And then pulling out an old cowboy's hat replete with dark face and a striped cane which he used as a pointer when it wasn't being licked or leaned on the doorman went on I mean ef you got de blood o de lamb on yo hans man how yo gonna look dis cow in de eye? Tell me dat!

Bos could read. That is surprising since so few ungulates can do that. And she does read which is even more. But Bos has had a reasonably good education. She has been to school. With John Skelton and there was a time when she knew Philip Sparrow well.[10] Bos dined once at All Souls a fine table in a room designed by Wren and she asked the Warden or was he the Master or the Head or the President she never could keep those labels pasted on the proper faces who reminded her of a frigate bird with long eye lashes what he thought of Philip and he said well I don't think.[11] Bos thought they were not related. Bos had a good feeling for Marvell too. Andrew understood. The wanton troopers riding by had shot an ungulate. Ev'n Beasts must with justice be slain.[12] Sometimes the ambiguity of that line bothered her even when she tried not to let it. Had Andrew noticed it and let it stand even so? She stopped to wonder about that. But Christopher Smart was a favorite poet *Jubilate Agno* the work and she would cite with approval

> For Painting is a species of idolatry, tho'
> not so gross as statuary.
> For it is not good to look with [y]earning
> upon any dead work.
> For by so doing something is lost in the spirit
> & given from life to death.
> For BULL in the first place is the word of
> Almighty God.
> For he is a creature of infinite magnitude in
> the height.
> For there is the model of every beast of the
> field in the height.
> For they are blessed intelligences & all angels
> of the living God.[13]

Do you think he demanded facing the group ringing him in the lobby which he did by slowly spinning round like a figure skater pivoting on one foot pointing his cane as he spun do you really believe that any herpetophobe can appreciate Lawrence's feeling for his snake? Lawrence had no snake muttered wired eyes. DH not TE explained the dirty jeans a poem about a snake. Well why didn't he say so? They're always going out of their way to be obscure. NOT ME MAN yelled the doorman evidently overhearing not overlooking the remark. We be speakin as plain an unfancy as can be. As Mr. Sweeney said I've gotta use words when I talk to you.[14]

You don't understand the doorman said simmering down. Sobering up discourse about bull fighting is doubly unwanted here. For not only does it betoken it establishes and that conclusively a lack of proper of right and fitting feeling not only does it testify to an unaesthetic and anti aesthetic insensitivity but it invites and welcomes indeed compells an altogether a wholly a completely invidious comparison. For mention bull fighting breathe the phrase and at once one's eyes are filled with and heads are bounded by violent images Goya's bulls are pawing the ground snorting charging dancing everywhere.

Reality and time reality he said is so difficult. Goya's bulls are real very real very very real if reality will bear an intensifier and language cooperates. Goya's bulls are so real and so is Bos of course in a way but not the same way. And it does matter and it doesn't.

It doesn't exactly matter if there is no cow that his picture is a picture of. Because if there isn't then there's nothing at all to know about it that is going to make any difference to the way you and this cow stand to one another. Because for example if there is no cow then no pinbone is misplaced and no dewclaw can be awry. If there is no cow her pasterns her hocks her crops her heart girth don't precisely matter brisket rump teats and chine don't in every detail matter. Cow doesn't precisely matter if there is no cow that is this cow here.

But of course cows must matter even so. Certainly if there is somewhere this cow. If anyone wants to understand Van Gogh's trees it is useful to go mad under the trees around Arles that he did to run and to glare to sit under cypress and look at the sun. If Leonardo's Lisa were here would you object to such a glass on that smile? Conceivably we have mistaken herniated tissue for psychological insight. If real Jesus were here he could be our window to look through to that gory twisted Grünewald's Christ. If there is somewhere a cow that cow could be a lens through which to view this cow with the subtile nose.

Is there a cow that is this cow here? Christ knows! We can do no better than wonder and then spit out the thought. This may be done by grasping the extremity of the tongue between the left thumb and left forefinger pulling gently but firmly till the fleshy process protrudes sufficiently to allow a careful thorough cleansing of the exposed epithelium. This may be done by scraping the exposed surface with the prepared nails of the right hand. How the latter are prepared said the doorman may be left for another time. For thought dissolves to spittle leaks out at the mouth.

To be told that the painting is oil and duco on canvas that won't mean too much to people who don't paint. If you do paint and you have the canvas at the tip of your fingers then knowing that how did he get that texture? What medium was used? A heavy paste? Something with mastic varnish and thickened linseed oil? A special sort yes. What about the cow with the subtile nose? First look at it and first look at the texture the surface the quality of the pigment. If one knew nothing about Dubuffet about his other works earlier and later one would still look to the texture of this painting. Because it is striking and remarkable. But knowing what one knows about earlier and later about his texturology about the beard map and beard paintings and if one had looked at his lithographs at his *Vie diffuse* at his *Grâces teńebreuses* one would look to the texture. Isn't that so with all paintings? No. Not with a Botticelli. He could have spit on the canvas. It wouldn't have made much difference. Of course his paintings are great paintings beautiful paintings he is a great draftsman. But pigment isn't anything he ever thought about. Dubuffet thinks about pigment about texture about material Dubuffet attends to all these matters and makes anyone who looks at a Dubuffet attend to such matters. So in looking at the cow with the subtile nose look to its surface first of all.

Good you're looking at its surface. Its surface is not smooth it is bumpy rough uneven. Right. So what? Surfaces of different things are different so what? Right. You're looking at the surface. Touch it! No don't touch it you can't do that if everyone were to do that we'd have no painting left pretty soon and in this city everyone would do it. No. Just look at it. What about it?

Imagine touching it! Yes. OK. You're imagining yourself touching it it's not smooth it's rough it's bumpy it's uneven so what? Well nothing that isn't the right idea. Turn your head around. That's not the right conception of a surface. It is the conception of the surface it is the conception that to begin with matters most. Of course it's either smooth or not smooth rough or not even or not and you can see that anyone can see that that doesn't matter very much. See the surface see it for what it is. Yes right. You can describe it. But that's not what is important either. What's important then?

Here you have a meadow mouse the size of a child's thumb. He sits in the palm of your hand. Think of that.[15] He's smooth he's soft. You can see that without touching him. But touching him is something else. Is that interesting? Put your hand to things. Let something be at your finger tips. Dog's breath a dripping tap a concrete wall old blackened smoothed by time. Perhaps if you go this way something will happen.

A description need not distinguish. It may fail to capture. Words may reach out but there's no guarantee they'll touch. This cow this surface of the Dubuffet it's rough it's uneven it's jagged it's pigment piled up. That doesn't seem to say anything. But this surface here and this canvas now this jaggedness this roughness this uneveness that's something different. Look at it!

There is the doorman interjected a philosophical problem here which can be alluded to glanced at but must certainly be passed by. For its adequate consideration would force an irregular attention a distraction upon us. Still the thought is bound to be there huddled in some corner of the head. Sitting rocking back and forth inducing a mild sense of imbalance a hint of vertigo. Precisely how precisely what is the fit between sight and touch? To feel smooth and to look rough that is possible. What are the possibilities and how are they exploited? And then will these possibilities remain once the eye and hands are trained? Are the eyes and hands of a technician assured of a complete and perfect correspondence between visual and tactile aspects?

Bos stood in the book store forgetting herself where she was what she was up to. She stood in a long long pause caught in time like glass. Looking and looking and looking at what? A book store clerk recovering from his astonishment stood sucking his lips running a pink verging on beige tongue across back and forth alternately licking and drying a tight drawn mouth. One could hear the words welling up from some remote sequestered area of his body possibly a cavity adjacent to the colon whistling through his esophagus to burst out at the teeth.

Look what you've done to the floor! An iridescent blue blowfly hums in her ear settles on her nose. She crosses her eyes in an effort to focus on him. A sense of urgency grips her muzzle. Look what you've done! *MMMOOOOooooo*. The startled approaching clerk leaping backward tripped over the feet of a pair of tourists. Crashing down at the four feet of the touring pair the impact of the clerk's skull against the ersatz marble floor was filtered by a lucky firm mound.

Bos was startled by the unseemly show ungulates having an awkward albeit delicate sense of decorum fitness and so forth. Not knowing precisely what to do about the reclining manure pated clerk displayed on the floor before her she decided not to do anything about it or him. She satisfied here innate requirements imposed on her through a minute sense of fitness by selecting and carefully fitting and carefully putting on an exotic look the look being deemed by her sufficient to the demands of the moment.

Bos remembered an artist once at once forgetting the floored clerk. He the artist sat with his head uncovered on a congoleum floor his works his pots of paint displayed for her appreciation as she stood posing in the nude. He sat working in an enormous bare room under a harsh white light an uncovered bulb fixed in a ceiling fixture an elaborate structure of tinted crystal multifaceted pendants designed to bear some thirty incandescent flares. His bald head shone glistened everywhere coated with sweat. Reflecting electrically it supplied unbelieving eyes. Bathed in the radiant flux he sat she thought wondering she wondered. There was after all not much that she could have done even if some notion of doing it whatever it might have been had somehow occurred to her. But nothing intruded on her consciousness no thought forced her entered her precincts of concern. She felt easy and free. Leaping clerks were not her affair. The disorder her presence seemed to have occasioned signified nothing. For disorder is a function of time a display of entropy. Time is nothing to Bos. Bos was frisky. The books in the store on the shelves propped against the walls did not intrigue her. She stepped over counters walked along strolled into the picture galleries.

She is standing looking. Time passes and will. That doesn't matter. What is important is this. What is slow and what is fast is a function of attention and of detail. We do not all of us process at the same rate and pace or with the same vivacity and flow or urgency and need. We do not equal one another in voracity in our appetite for time. We often fail to equal ourselves. We do not any one of us ever do any of these things in the same way over prolonged or even minimally extended periods. Who can appreciate Bos' pauses? Bos was forever stopping. This of course seemed not at all unusual to her peers. Indeed were they somehow called upon to testify to their existence the pauses of course quite possibly they would have been at a loss to do so.

Most of us move between long pauses. Bound by day to night by waking to sleeping Bos worried about that too. We must sleep. But I am consoled by this. Pink flamingoes sleep. Not that I truly understand this matter. They sleep yes that is clear to me. I recall a long summer each week watching pink flamingoes sleep. They sleep on one leg the other bent tucked up nicely wing leg head and long neck all neatly together. Some sleep on their left leg some on their right. Bos wanted to know. Did those that slept on their left leg one time did they always? Are some right legged sleepers and some left legged sleepers? We didn't know. And the truth of the matter is though I am a firm admirer of flamingoes one does look like another. So it was hard to say. Bos thought to subject the problem to a crucial experiment. She would adopt the stance of a flamingo and then see whether a right or a left positioned posture would readily assume the character of a habit. But she was put off by the matter of standing in a pond.

It is consoling that pink flamingoes sleep. There is no anxiety to their sleep. It is apt and everywhere equal. An harmonious balanced pause a just interval between hectic displays. A lovely stillness between flights of fishing between writhing elongated necks intent on grooming and a sudden nervous flutter the long light flashing breaking the surface of their familiar waterways.

Bos stands as a wide awake pink flamingo in the picture gallery. Not a one legged stance but balanced lightly on all four at peace a long soft thoughtful pause. Looking.

4

Antiart was Dada a long time ago. What Dada means the name isn't too interesting. Yes yes from Rumanian *da da* as in Russian *da da* or hobbyhorse which is what *da da* means in French. One needn't worry about that. Dada was a movement an event. Around 1915 maybe 14 16 18 17 various dates in Paris Zurich Geneva Berlin. It doesn't really matter where. Dada has entered art history has ceased to be. That doesn't mean that antiart has ceased to be. (Antiart is one word as I use it. Antiart. Antiart is good. Antiart as I use it is not anti art. To be anti art is bad.) That Dada has ceased to be does not mean that antiart has ceased to be or if it does it need not and anyway and even more it could be good if it did not. For a conception of antiart could enrich a conception of art. So rather than thinking that antiart was Dada and like Dada is gone I shall think that Dada was antiart and all that is gone is Dada an historic manifestation exemplification instantiation of antiart.

Is there today anything that is antiart? What about readymades? (You know what readymades are.) What about garbage cans? What about an artist who confides that his great work of art is this. He woke at 6:25 a.m. he didn't say anything he just thought something. He then went back to sleep. Do such things occur? Yes they occur. Is that a work of antiart? How far can one anyone go and so on? And what makes it what it is and and and and and so on and on? I don't know why it is so easy much too easy to ask the wrong questions when one thinks about art. It is and they certainly are.

An evolution of a conception of antiart from that of art is or anyway could be at least analogous to that evolution and conceptual expansion exhibited in the current conception of antimatter. There is an analogy only analogy. The analogy is suggestive. When Dirac's wave equation for the electron was solved physicists were surprised to find that the solution yielded both a negative and a positive frequency.

Since frequency in quantum mechanics is proportional to energy it was hard to see what the negative answer could mean.[16] Dirac was able to prove that the negative frequency corresponded to an electron with a positive charge what was later called a positron and antimatter. A surprise certainly. An expansion of a physicist's conceptual scheme. And when some time later Anderson (Carl D. of Cal Tech) discovered the positron it became clear that the expanded scheme had not lost contact with reality. All this is only an analogy. There is no need to stare it out of countenance. Art has no relevant equations. There are however factors that one can focus on. Aspects of works. Dimensions.

Let's be old-fashioned very much and think of beauty. And suppose what isn't exactly so so that beauty has its opposite and that that is ugliness. So I mean to deal now not in analogy but in figure. This needn't worry anyone overmuch. It is not as though we had set-theoretic matters to set straight. Let's suppose that there is a set which is the set of beautiful objects and this set has as its complement the set of ugly objects the set of objects that are not beautiful. Some artists have in past and by their works explored the set of beautiful objects. That is an easy figure that I am asking you to entertain. I am supposing that Poussin Raphael Michelangelo and so on have all explored this set at great length. Not *ad infinitum* and certainly not *ad nauseam* but certainly at great length though *ad nauseam* for some people. Perugino was a master in the exploration of the set of beautiful objects. An artist whose work constitutes an exploration of the set which is the complement of the set of beautiful objects the set of ugly objects could then be said to be in this respect an antiartist. His work could be said to be works of antiart.

Is this right? Think of Dubuffet. Isn't he an artist and not an antiartist? His work has been in part anyway devoted to an exploration of the set of nonbeautiful objects. I'll show you what he has said.

> The idea that there are beautiful objects and ugly objects, people endowed with beauty and others who cannot claim it, has surely no other foundation than convention – old poppycock – and I declare that convention unhealthy. I would like people to look at my work as an enterprise for the rehabilitation of scorned values.[17]

Dubuffet is explicit in rejecting the distinction between the set of beautiful and the set of nonbeautiful objects. Yet in his own work he explores or could be said to explore primarily the set of nonbeautiful objects. In that respect he could well be classed an antiartist and his work could be classed works of antiart. Certainly his cows are not beautiful. In some sense or other on the other hand they are very beautiful but not beautiful as is a Poisson or Perugino or Raphael or anything like that. Very very beautiful in a different way. His work constitutes an exploration of a negative aspect associated with a negative characteristic of art. And in so far as it does constitute such an exploration to that extent I would suppose that one could characterize his work as works of antiart. But yesterday's antiart can be and here is today's art and Dubuffet is an artist his works works of art.

Obviously what interests philosophers theoreticians aestheticians beauticians and the like is how far one can go in this direction because it is easily obvious to almost anyone (who can read me) that contemporary art modern art of this century has thrived on the rejection of certain aspects of previous art. If we took for example not the aspect of beauty but that of representation better yet pictorial representation then it is very clear that with the development of modern abstract and nonobjective art with respect to easel painting one encounters non(pictorially)representational works. It won't do to say that nonrepresentational art developed in this century. That's untrue. Some art of some ancients Islamic art generally was nonrepresentational because of a taboo on representation in connection with religious objects. Marvelous nonrepresentational pavements have long been underfoot in the cathedrals of Europe. It isn't that nonrepresentational art suddenly appeared in this century. What did materialize was nonrepresentational art in connection with pictorial works easel paintings and the like. They had never existed before. No one as far as I know did nonrepresentational easel paintings prior to this century. Nonrepresentational work had always been decorative in character whereas modern abstract and nonobjective paintings are not decorative neither are they works of decoration in any sense nor are they even in a decorative tradition. So in contemporary art what has happened is that another particular aspect of past art has been negated. One can list aspects that have been negated in this sense sets that have been explored. The set of nonbeautiful objects was explored considerably and this early on in the century. Think of the Fauves. And the Dadaists were about early on. The set of non(pictorially)-representative objects was also explored. Another aspect that one finds in modern art contemporary art even more surprising though in a way not too surprising is nonorder and the exploration of the set of objects that lack order. In a way this is not surprising because if one contrasts renaissance art and western art generally with oriental art the kind of order one finds in oriental art is radically different anyway so it's not surprising that in western art artists began to explore different types of order. For example the characteristic organization of a renaissance painting is on the diagonal from the lower left to the upper right. No such diagonal is likely to be found in a Chinese scroll painting. But when I speak of the rejection of order in contemporary painting that is not what I have in mind. I am not referring to the difference between oriental and western painting. I am talking rather say of the difference between the work of Poissin or Van der Weyden or Tintoretto or any of those people and say the work of Mark Tobey the work of DeKooning the work of Motherwell or better yet Pollock. In Pollock there is no organization whatever that is comparable to the organization in a characteristic Baroque painting. Not only is there no dominant diagonal or organization spatially three dimensionally it is generally an over-all pattern that is presented which has no distinct organization. In Mark Tobey's work very beautiful work one finds nothing like a dominant diagonal anywhere but all kinds of little things everywhere. If one were to draw an analogy between painting and music then Mark Tobey's work is quantitative rather than climactic in character thus akin to eastern and near-eastern rather than traditional western music.

So we have beauty representation organization all rejected in contemporary painting. What else could there be to accept or reject? Design pattern contour. Contour disappeared a long time ago in all kinds of painting. In fact the rejection of contour is the striking difference between the work of the Florentine and that of the Venetian painters. The dispute between the disciples of Ingres and those of Delacroix was precisely over the importance of contour and the associated qualities of draftsmanship. Whereas with Delacroix and his disciples and with the Venetians what mattered most were what are called the painterly qualities. What the Germans call *malerisch*. There's no good English word for it. Painterly is about the closest. Pigment becomes important. Picasso's analytical cubistic works are very strong in these painterly qualities. He had a feeling for pigment. Titian's are great in this respect. The pigment is magnificent. But Botticelli might just as well have used spit to paint with I mean there's no pigment it's all pure line and yet he's an angel. He's (don't say was) is a great draftsman. Ingres too. Yet Ingres gave no sensuous quality to his pigment. Delacroix and his school thought that these painterly qualities were absolutely essential. Ingres and his school thought that contour was absolutely essential. A good critic like Baudelaire said they were both very interesting. A very sensible man. It's clear that art can evolve in either direction. German expressionists strongly emphasized the painterly qualities deemphasized Botticelli-type or Ingres-type contours. German expressionists really had no interest in draftsmanship though with Beckmann and people like that one finds very powerful lines but it is not the careful contour of Ingres but something else altogether.

These are obvious aspects of works of art representation beauty organization quality of draftsmanship painterly qualities. Space a study of space this would again be something rejected by some painters not by others. In Cezanne one finds an enormous stress on the picture plane and spatial properties. In contemporary action painting it is totally discounted it is totally irrelevant. The question which seems to interest aestheticians which I think is a very bad question a stupid question in a way but it is one that some are inevitably driven to is this. How far can you go? In this direction piling up the list of complementary qualities that are being explored. If you take not only works which have no beauty but are not pictorially representative disdain order have no contours of any kind draftsmanship has been thrown out the painterly qualities are totally rejected. Let's say we add to our list chromatic things like color Albers' works studies in color and spatial relationships of colors we throw that out too. We throw out any interest in space whatever that goes too. How far can you go in this? An extreme case is what is wanted. A work which for a moment anyway at this point in time would look like it the ultimate in this respect. A work that rejects all the qualities we can think of. I'll call it for very good reasons the *mu* work. *Mu* in honor of Joshu a zen master whose favorite form of zen was *mu*. Which means nothing. Literally nothing n o t h i n g.

Suppose we have an artist who reveals confides that his great work of art is this. He woke at 6 : 25 a.m. he didn't say anything he just thought *mu*. He then went back to sleep. That was his great work. Is that a work of antiart? It is not yet a work of art. That is clear enough. Is it a work of antiart? Doesn't it fit into the categorization I have supplied because it constitutes an exploration of the complement of a rather complex set that set constituted by the complement of the union of the various sets we have been considering? This work then is equivalent to the utmost and most unmistakable work of antiart. It is as far out as anything can be (or so it seems to beauticians of the moment). Could this be a work of antiart and is antiart of any interest to us? That I take to be the fundamental question of interest to theoreticians and beauticians concerned with antiart. I can't think of any other question that is really of any significance. And I think the right answer to this is rather strange. I think the right answer to the question whether or not *mu* this work *mu* of the artist who wakes in the morning at 6:25 and thinks *mu* whether that is a work of antiart the right answer to that is probably of the same character as the work itself. That seems to me the most reasonable thing one could say about it. That if you want to say whether or not it is a work of antiart hence soon or eventually of art in any form or other you probably have to say if you want to be reasonable about this that that of course depends on the work. And there isn't any better answer to give.

A long time ago Bergson in one of his essays said that he could not tell you what the art of the future would be like because if he could tell you that he would do it himself and then it wouldn't be the art of the future. He would produce it at the time. And I recall my old teacher Arthur Murphy telling me that that's a really dumb answer because he Arthur Murphy could tell you what this chicken was going to do namely lay an egg but he Arthur Murphy could not lay an egg. It's quite one thing to be able to predict what is going to be done and quite another to do it. So Bergson's argument that if he knew what was going to happen he could do it seems to me very bad. Nonetheless there is a sense in which if you really were in a position to predict and so it would seem in this case to create this work of antiart then it still would not be and is not clear whether or not it would become and so eventually be a work of art. Let me try to explain that because this is the main thing I have to tell you that is of any interest.

I'll put it very plainly. I don't know if I will ultimately put it so plainly but I'll put it very plainly for the moment. Whether or not *mu* is a work of art or a work of antiart and hence sooner or later of art because I take antiart to be a revolutionary manifestation of art whether or not *mu* is a work of art depends on *mu*. And if you say well I have just told you what *mu* is then my answer to that is that

I haven't told you what *mu* is though I have told you what *mu* would be like. Namely an artist waking at 6 : 25 a.m. thinking *mu* and going back to sleep. And that doesn't sound like a work of art. I certainly grant that. That sounds like utter gibberish. Or if it is a work of art I couldn't care less. It's like the dinner you cooked for me when I wasn't there. You described this great dish but one of the specifications was that I wasn't there. I'm not going to argue about the taste of it the quality and so forth. If I wasn't there the hell with it. Now you might say the same about *mu*. But it's not clear to me that *mu* isn't a work of art.

Let's consider. How could *mu* be a work of art? It seems absurd doesn't it to suppose that *mu* is a work of art? *Mu* seems as unlike a work of art as anything could be. It seems very much like antiart when antiart seems like a lot of nonsense and if that is what antiart is then this is nonsense and so is antiart. Whereas I am inclined to think that *mu* might in fact be a work of art. I think that is a real genuine possibility. How could that be?

One way it could be is that there could be a setting for it. Which could account for it. Let me explain what I mean by a setting for it. There is a strong inclination to think that because I have taken the negation the complement of a complex set that that somehow characterizes what's in question. I've told you that it's not this not that not that and so forth and you think you have a clear idea of what this is. I've also described to you a man who in the morning at 6 : 25 wakes in his bed thinks *mu* and that's his work of art. That sounds like I have really described everything to you. I haven't. That's where you go wrong. I'll give you an example that will strike a responsive chord in some (those familiar with the paper I refer to). I say a cheetah can outrun a man and you think you understand that. But if you start thinking about it you can see that there are a lot of troubles.[18] Because I haven't told you which cheetah I'm talking about and about a weight on his back or whether his legs are broken or under what conditions the race will be run and so forth. Or I tell you that you don't have to be afraid of lions. Why? Because the last free one has been locked up in a zoo. There are always all kinds of special conditions. You think you understand what is said but if you start bringing in the supplementary conditions you begin to see that things are rather different. The idea of a setting or of supplementary conditions is a very important idea in the philosophy of science. Anybody who does philosophy of science or any physicist knows that there really are no predictions that you can sensibly make about say the behavior of planets in our solar system unless you have some specification of supplementary conditions. For example suppose you say that you've seen celestial mechanics and updated general relativity theory and you're in a position to make a very good statement about the perihelion of Mercury. And I say well last week somebody sighted a comet and it was in the neighborhood of Mercury and what do you say to that? Are you going to throw out your newly revised Einsteinian relativity theory? No. What about your predictions? You will modify your predictions in the light of this newly observed comet. Because the new comet isn't a feature of your theory. The comet is a novel momentary aberration which the theory doesn't pretend to cover and until you get some good data on the comet on the position of the planets and so forth and how it's all going to tie together you don't make many predictions. And if your predictions turn out to be wrong because there was a crazy comet shaking its tail that doesn't refute the theory at all. The theory isn't concerned with such supplementary matters. Just so you can't refute me when I say that strychnine will kill you dead by saying there is this woman in Harlem and she takes strychnine all the time and it doesn't kill her. Because I say she's developed an immunity she spent the last forty years of her life building up an immunity. So what? That doesn't show that strychnine isn't a deadly poison. Supplementary conditions are brought into every specification of a scientific event. I want to require the same thing here. You say because of how I described it that *mu* isn't a work of art or of antiart and so eventually of art. How do you know that? What makes you so sure it isn't?

Let me remind you of things that couldn't possibly be works of art that were works of art that proved to be works of art. Years ago if you had put the question to somebody do you think that the seat of a bicycle and the handlebars of a bicycle stuck together could be a work of art. How foolish can you be? How could that be a work of art? Handlebars of a bicycle and the seat of a bicycle. Nonsense. Picasso's great bull is a great work of art is just that. Just about anyone sensitive to art sees that that is a tremendous work. Do you think that something that is virtually identical to a configuration of mud could be a great work? I mean it looks like mud. If you saw just sections of it you'd think it was mud. How can that be a great work of art? Well there's mud and there are these paintings by Dubuffet. They cost 30 40 50 thousand dollars. Rightly. They're great works of art. They look like mud if you just look at them that way. Described in that way they are nothing. That's the wrong way to describe them.

To take a more extreme case one supposes if one's rational that there's something which constitutes an answer to a set question. That is if you formulate a question which is of a certain form *q* then you think this can somehow give rise to something of the form *a* which is an answer to the question and by and large one is inclined to think that there can be an adequate characterization of what an answer to a question is. By doing a good syntactic and semantic analysis of the explicit symbolic or linguistic form of *q* you'll find what *a* is or can be expected to look like. That's a very common notion. So people begin to think that they can recognize an answer to a question by looking at the question and looking at the answer. Whereas it is fairly clear that if one turns to very sophisticated matters that one can't do this. What counts as an answer is something that you learn to count as an answer which is not to say that it is arbitrary or conventional or anything idiotic like that. For example if the problem is a problem in one-upmanship as in Potter's *Gamesmanship* then what counts as and constitutes a one-up answer to a question is a really deep and subtle affair. Only someone with a real flair for it can do it.

If one enters an even more sophisticated area of human discourse namely zen question and zen answers zen koans then one realizes very quickly that what counts as an adequate answer to a zen question cannot be characterized simply in syntactic or semantic terms nonetheless there are correct and incorrect answers to the questions. Now I am inclined to think that when it comes to the character of something an event an object or what have you as a work of art or as a work of antiart that there really is no reason to say that my *mu* can't be a work of art of antiart. It's true as I've characterized it it lacks every quality that we're interested in has no quality that would concern us. But that doesn't prove anything. So let me give you one final example which will illustrate the point I have in mind. Of course this may not persuade anyone. It persuades me but it may not persuade you. This is a problem that was put to Joshu a zen master. There was a dispute in a monastery over a cat. The monks of the right wing wished to take care of the cat. The monks of the left wing wished to take care of the cat. There was controversy. To end it the abbot said that unless someone could resolve the issue he would cut the cat in half lengthwise giving half to the left wing half to the right. Joshu was in the fields away and no one spoke and the cat was divided. When Joshu returned the monks gathered round him and put the question to him. If he had been there what then? And Joshu gave what I take and deem to be the best of the known zen answers. He stood for a moment. He took off his sandals put them on his head turned around and walked out.

There are some koans which do not have answers. Let me give you two of those. These are problems now there are no known answers. Zen masters come up with suggestions but there are no recorded answers that I know of. Two problems. One. Your feet and hands are bound. You are hanging over a precipice clutching a reed with your teeth. A stranger passing by leans over the cliff and asks you what is the meaning of zen? That is a standard zen question. What do you say or do? You are starving lost in a desert and you find a sacred bottle with a tiny neck filled with an unsacred goose. You may eat the goose could you get to it but you may not break the bottle. What do you do? Now what counts as an answer to these koans? The question what counts as an answer to these koans is strikingly analogous to the question what counts as a work of antiart when you go way out when it's as far out as when you've negated all the familiar qualities. No answer to these koans will do that is of the following sort. I poured water into the sacred bottle cooked the goose made soup out of him poured him out. If you say that the proper reply from a zen monk might be to whack you with a stick cut off your finger boot you back to your meditation pad unless he congratulated you on your unusual perspicacity. I mean you've had it that's no good at all. If you say in the other case when you're clutching the reed with your teeth help. No good. You've had it again. That's a lousy answer. So those would not be correct answers and if you think of *mu* along those lines as just an artist waking in his pad reeking of garlic sweat just thinking *mu* drowsily and then going back to sleep that's a work of art? That's terrible. Just an idiot mumbling to himself. That's not a work of antiart at all. But the fact that that isn't a work of art doesn't mean that there isn't some work of art having almost the identical form. What counts as good answers to these two koans? That's not written. And since I'm not a zen master I can't give you a good answer. I can easily give you answers which are as bad as the answers I just gave you. Answers which are as bad as the ones I gave but which wouldn't be as bad as them if they weren't being given here by way of example. That of course makes everything very different and difficult. So another bad answer is that I polished the sacred bottle with the goose down picked my teeth with a quill. That's a terrible answer just as bad as the other answers no it wouldn't be just as bad if it weren't coming up the way it's coming up. If it were in answer say to some other question. But coming up the way it's coming up in just this position in our discourse it's just as bad. And the other one where I'm or you're clutching the reed with our teeth a very bad answer but now this time not as bad an answer maybe it's the best so far is will you tell me again what the problem is? That's not too bad. It's more in the right direction. But it's still no good particularly in this context. We should not expect too much of ourselves in these matters.

If you suppose that there is no correct zen answer because there is no specific characterization of the conditions to be satisfied by an answer then I think that you are making the same kind of mistake you are making when you suppose that *mu* can't be a work of art. I recall hearing a philosopher say that Wittgenstein and zen have a great deal in common because both Wittgenstein and zen masters reject silly questions. Wittgenstein would brush off silly questions. Like how do I say today is Friday meaning that today is Saturday? Go ahead and mean that today is Saturday but say that today is Friday. That's a silly thing to ask and there's no way of doing it. But the philosopher was somewhat naive because he supposed zen masters give silly answers to silly questions and only silly questions whereas in fact from his point of view they give silly answers to all questions. Any question is going to be rejected. There is nothing that counts as a straightforward answer from a zen point of view. Any question put by a zen master if he's himself won't have a straight answer unless of course what is really deviant is a straight answer in which case he'll give you a straight answer unless he doesn't. It's always a move in a very sophisticated enterprise.

Mu I'm saying this work of antiart and so eventually of art may have that peculiar character. Just as a zen answer is an antianswer and yet has a specific but as far as I can tell unanalyzable quality which makes it a correct zen answer. Which practitioners of the religion can identify. And which one can begin to get a feel for. Just so *mu* could have this a specific quality. And though my description of *mu* makes it utterly boring banal uninteresting that description hardly exhausts the potentialities of the situation. There could be something corresponding exactly to that description and yet having a certain ineluctable quality. And for me or us to pronounce now that no it could not have any and there could not be any such quality would be an egregious piece of folly. The development of aesthetic practice and scientific theory keeps surprising us. We keep discovering aspects of experience which we never expected. Certainly the discovery of an electron with a positive charge was astonishing. No one had suspected that there could be any such thing. And indeed the current suggestion that perhaps nothing is rightly classed an elementary particle that perhaps such entities are better thought of as stages in the transformation of energy such things are altogether surprising beyond the imagination of anyone when particles were first spoken of. I see no reason in art to inveigh against the possibility of the total rejection of present aspects of art.

Is *mu* to be a work of art by being a work of antiart? Anyway since yesterday's antiart is after the revolution today's art if we wait long enough the question will run off into the past wagging its tale behind it.

II

THE COW ON THE ROOF

On hearing this

I think: that's Milhaud's *The bull on the roof (Le boeuf sur le toit)*. Is it? Maybe it's *The cow on the roof* and not *The bull*. Those two sound much alike. *The bull* is being performed in New York we're in Atlanta. That doesn't matter. A piece of music doesn't have a spatio-temporal location. *The bull* isn't ever either here or there not even when it was being composed in Paris in 1919. A piece of music is an auditory pattern. An actual performance of a piece of music is a physical realization of a particular auditory pattern. Which auditory pattern is being given a physical realization here? Is it *The bull* or *The cow*? It sounds like *The bull* to me. It sounds like *The cow* to me. That's because they sound alike.

But listen to that: that can't be *The bull* there's no such chord in *The bull*. But then there's no such chord in *The cow* either. So what can it be? It's one or the other that chord doesn't matter. That was just a mistake. A flute player was attacked by a bee he sounded a wrong note. That's all. How do I know he sounded a wrong note?

Maybe it was the right note and it's not *The bull* just a different work. Not *The cow* either perhaps. Why do I say it was a wrong note? If it was the right note then there's something wrong with the piece of music. This sort of alternative confronts one at every turn. This kind of question stares out of every corner. Is that painting hung upside down or is it a badly unbalanced work? Why can't I make that move? Because it's strategically absurd or because it's illegal? Is he cheating or playing according to different rules? How does one tell whether a wrong note was sounded?

That's a delicate question. One listens to a lot of music. One acquaints oneself with the work and the tradition it's in: after a while one can tell. It's not simple but it's not hard for humans at least those that are sensitive to music. We hear this

and we say no the E^{b} is a mistake: melodically harmonically it is completely out of character for Telemann. Besides one knows why the note was flatted: the fiddler shifting back from 3rd to 1st position overslid. It's more plausible to suppose a fiddler's finger slipped than to suppose a remarkable failure of Telemann's ear. But how does one tell whether a wrong note has been sounded?

Is this a simple matter? I suppose that depends on how one looks at it. Certainly it's a simple matter if what's in question is can people do it without much difficulty because they can. It's not a simple matter certainly not simple at all when one has in mind to analyze to detail the relevant factors. The E^{b} is evidently a mistake and it's simple enough for any person sensitive to music to hear at once that it's a mistake. But the factors involved are enormously complex. Yes do not underestimate this. An explicit analysis would call for the discrimination of various subtle highly theoretic factors. A conception of melodic structure and melodic line is called for. A conception of harmony and so of tonality is called for. Some conception of bowing is relevant. A theory or the like about the character of a composer's abilities is also relevant here. How does one tell whether a wrong note has been sounded? I'd call this a theoretic question were it not for the fact that no single theory short of a super-theory of human behavior could hope to encompass such disparate factors. But how does one tell whether a wrong note has been sounded?

Once I did a drawing of what seemed to be a cube and someone looking at it said your perspective is off your perspective is wrong and I said no the perspective is true that's a perfectly accurate drawing of an odd shaped object.

If you know the piece of music if you've heard it before then you might very well know that a wrong note has been sounded. And there the test is simple enough. Does what you hear now conform to what you heard before? But maybe before when you heard it a wrong note was sounded and now when you hear it what you take to be a wrong note is the right note. Perhaps the lack of conformity between what you hear now and what you heard is owing to a fault having been committed before and not now. All that's clear now if your memory is correct is that what you hear now is not the same as what you heard before. That doesn't tell you which is the correct one if either is the correct one. How does one tell whether a wrong note has been sounded?

Maybe sometimes there's nothing to tell. It might be that sometimes this is just a matter of making up one's mind which way to take the work. If a wrong note has been sounded then this is not the way the work sounds and if it hasn't then perhaps this is. Perhaps in some cases we don't tell that a wrong note has been sounded: we decide that it has been sounded. And perhaps it is difficult to tell when we're telling and when we're deciding. How does one tell whether a wrong note has been sounded?

One might look to the score. Scores can pose problems much as performances can pose problems. A piece of music is an auditory pattern and a score is a representation of such an auditory pattern.

Scores may have wrong notes as performances may have wrong notes. So the same questions come back and back again when one turns from performances to scores. No autograph copy of Bach's suites for the violoncello has ever been found. The Bach-Gesellschaft edition is based on a manuscript written by Anna Magdalena Bach's second wife and on another written by Johann Peter Kellner one of Bach's contemporaries. It also uses some of the earliest published editions. But these scores differ at certain points. Anyway even if one had an autograph copy of the suites one knows perfectly well that even autograph copies may contain errors. (What could be called the lower criticism of musical scores is no less perplexing than the higher.) On occasion an error in a score can easily be seen to be an error if it's sufficiently glaring. But on occasion it's altogether unclear whether what seems to be an error is an error. In adapting and editing for viola the first violoncello suite Lifschey notes with respect to the twenty-sixth measure that while there is no accidental before the second *B* of the measure either in the Anna Magdalena manuscript or in that supplied by Kellner B^b appears in all the early editions. Lifschey in his adaptation makes the *B* natural. Some performers choose to make the *B* flat. Which is correct? Lifschey thought evidently that *B* would do very well since it's what he supplies but B^b does well enough which is what some performers play. The Bach-Gesellschaft edition makes the flat optional. Neither choice poses any harmonic problems. Melodically the augmented second supplied by B^b was not implausible for Bach. Which is correct? Evidently either will do but what Bach wrote is not known.

Wrong notes and scores don't matter. What we're hearing is *The cow* and not *The bull* that's clear enough. What makes that so clear? It's fairly simple. I heard two oboes. *The bull* is not scored for two oboes. It's scored for only one oboe. So since what we're hearing has two oboes in it it can't be *The bull*. It can't be *The cow* either if that's right because *The cow* is scored for only one oboe. Then it's neither. No that doesn't follow at all. There's no reason here why one can't double up the oboes if one wants to. That doesn't make it a different piece of music. How about three oboes then? All right. How about four then? All right. Five? Six? Seven? All right. How about fifty oboes? Well .... How about ten thousand oboes? No. What does that prove? Nothing. Not even that one can have a field day with the fallacy of the sorites. We knew that already. A piece of music is an auditory pattern. The number and timbre of the instruments that the work is scored for is certainly sometimes an important aspect of that pattern.

One can within limits tamper with the timbre of a work but one can easily go too far. Debussy's quartet arranged for kazoos and guitars would be silly. But Bach's *Art of the fugue* is readily performed on an organ or by a string quartet or even by a small orchestra. There are limits to how far one can go in tampering with the timbre of a work but there is I think no way of defining these limits. The timbre of the individual instruments sometimes doesn't matter overmuch and sometimes it's crucial. Works exemplifying Hindemith's concept of *Gebrauchsmusik* lend themselves to a free and easy interchange of instruments. But the Debussy quartet hardly qualifies as an instance of *Gebrauchsmusik*. Not all violins violas and celli are alike in their tone in their timbre. There are some old Dutch violas with (marvelous) nasal tones: such a viola is not wanted in a performance of the Debussy quartet: the lush sensuous qualities of the music would be spoiled would be incapable of rendition on such an instrument. Sometimes a work scored for one instrument can be performed on another and sometimes not. These matters do not admit of generalization. One can perform Bach's *Goldberg variations* on a piano instead of on a harpsichord and all will be well. This might suggest that harpsichords and pianos are generally interchangeable but there is no truth in that. It all depends on the work in question. The Goldberg variations lend themselves to performance on a piano as well as on a harpsichord. But it would be an egregious error to play and who would really seriously attempt Prokofiev's sonatas for piano on a harpsichord? That these matters do not admit of generalization is not surprising. If you are surprised you should not be. Neither is it our inherent stupidity that precludes this possibility. New art new music create new interests and occasion the development of new means of production and expression. The modern piano is a percussion instrument: Prokofiev's sonatas exploit this aspect. Only if one could foresee all future developments in art and music could one hope to generalize here.

I think that what we're hearing is *The bull.* I think that what we're hearing is *The cow* not *The bull.* But that's just what *The bull* sounds like. That's just what *The cow* sounds like. Are you sure there's any difference between them? Maybe we're talking of one and the same work. *The bull* is *The bull* and *The cow* is different though it sounds very much like *The bull* there's no doubt about that. What's the difference between the two?

Consider the pair of sentences 'The sun's rays meet' and 'The sons raise meat'. These two sentences are two different sentences and yet the utterance of one may be acoustically indistinguishable from the utterance of the other. If we think of the utterance of a sentence as akin to a performance and of the sentence as the work performed then in this case with this pair of sentences given a performance one might be hard put to know which work was being performed. The (obvious) musical analogue of such a pair of sentences is to be found in a pair of enharmonic equivalents. Given our well-tempered scale the augmented second F–$G^{\#}$ sounds just like the minor third F–$A^{\flat}$ when played on a modern piano. However when these intervals are embedded in some appropriate harmonic context they are readily distinguishable. Is it that *The cow* is an enharmonic equivalent of *The bull*? That would be odd. *The bull* is primarily in C major. What key can *The cow* be in?

There seem to be at least two cases to consider. First the enharmonic equivalence may simply be a notational device devoid of any musical or acoustic significance. The 3rd Prelude and Fugue of

Bach's *Well-tempered clavier* is in $C^{\#}$ major but Franz Kroll in the *Peters* edition supplies an additional version in D^{b} major in the back of the volume: five flats are easier to read than seven sharps. In which case *The cow* and *The bull* would be one and the same work if that were the only difference between them. Do atonal composers attribute any significance to enharmonic equivalents? Notational convenience legibility ease of inscription these would seem to be the relevant factors governing the use of enharmonic equivalents in scores of atonal works.

Second the enharmonic equivalents may in fact be acoustically distinct in that there may be either a pitch difference or a difference in tone color. String players unlike piano players often supply a difference in pitch between $G^{\#}$ and A^{b} when $G^{\#}$ is a leading tone. But even in cases in which there is no change in pitch when enharmonic change indicates modulation performers sometimes supply a difference in tone color. Do *The bull* and *The cow* differ then in this way?

No that isn't it at all. There is no significant difference either in pitch or in tone color between them. Anyway a difference in pitch without a change of intervals doesn't matter all that much. If one night I hear a work which differs in pitch from a work I heard another night it doesn't follow that I heard two different works. If that were so then European orchestras would rarely perform the works that American orchestras perform since European orchestras often tune to *A* (above middle *C*) at 436 whereas American orchestras often tune to *A* at 440. What about *A* at 430? All right. How about 420? That's all right. Handel's tuning fork is 422.5. How about 410 400 390? That's all right. The great organ in Strasbourg Cathedral dating from 1713 has *A* at 393.2. The pitch of *A* (above middle *C*) has in fact ranged historically from 370 to 567.3. Frequencies don't matter all that much. It's not frequencies that one should attend to but intervals: intervals figure in scales. And intervals are fairly constant.

Why don't frequencies matter all that much? That's something to think about. Why don't actual pitches matter all that much? They certainly matter somewhat. There's no denying that. Then why do they matter at all? If there's going to be a performance of a piece of music then there'll have to be some frequency assigned to the notes and whatever frequency is assigned it will have to be such as to render the work audible and performable. So the frequencies certainly do matter in these ways. And if a composer composes with a certain frequency in mind then if one wishes to give a physical realization to the pattern that the composer created and if one wishes to do this without altering the character of the created pattern one will have at least to approximate to the frequencies that the composer presumably had in mind. So that's another reason why frequencies matter somewhat. But why don't they matter all that much? That's something to think about. The answer is easy but serves to emphasize the fact that pieces of music are auditory patterns. An auditory pattern may be realized at various frequencies. There is always some and often considerable latitude for the performer. So long as there is some systematic interpretation so long as the character of the intervals is not spoiled is not significantly altered and so long as high notes sound high and low notes low the created auditory pattern can have adequate physical realizations at various frequencies. (Any string player knows this. If one doesn't have absolute pitch and if no tuning device is available one may tune one's fiddle by tuning *A* at more or less of a guess and then adjusting the intervals to perfect fifths. One can then cheerfully perform without further ado.) (But why do high notes have to sound high and low notes low? Because if they don't then they won't be what they are. That's not illuminating. But it's true. What makes a high note high? I don't know and even if I knew would I be able to tell you?)

If *The bull* and *The cow* don't significantly differ in pitch or in tone color if they sound alike then aren't they one and the same work? That a performance of one doesn't significantly differ in pitch or in tone color from a performance of the other that the performance of one sounds much like a performance of the other or even exactly like a performance of the other that doesn't mean that they are one and the same work. But if there is also no question of one being an enharmonic equivalent of the other then aren't they one and the same work? Things needn't be that simple.

To speak of a piece of music and to distinguish between the piece of music and any actual performance is to speak in a remarkably abstract way. Not every aspect of a performance not everything heard at an actual performance is attributable to the music. If the performance was a disaster because half the audience had bronchitis one doesn't attribute the cacophony to the character of the music. Of course one can say that the coughing wasn't part of the performance and it wasn't. But that only indicates that when one speaks of a performance one is also speaking in an abstract way. And not even everything that is in fact part of a performance is to be attributed to the work being performed. If a violinist inadvertently scrapes with the wooden part of the bow though this occurs as part of the performance the scraping sounds are not to be attributed to the piece of music (given that such scraping is not called for in the score).

One aspect of a performance that is usually attributable to the piece of music is that of order. Pieces of music may be divided into parts and these parts are in an actual performance given a physical realization in some order. There is an actual temporal sequence in the realization and presentation of the parts. Beethoven's first quartet the *Opus 18 No. 3* is as was usual in four movements: *Allegro Andante con moto Allegro* and *Presto*. The work is generally performed in that order with a brief pause between each movement. If a quartet were unwittingly to perform the *Opus 18 No. 3* with the parts in a different order interchanging the second and third movements so that the actual sequence of presentation was *Allegro Allegro Andante con moto Presto* one would say that that was an incorrect performance of the *Opus 18 No. 3*. Certainly it was not what Beethoven indicated. That that is so is a matter of historical record. From a musical point of view I am inclined to suppose that there wouldn't be anything particularly wrong with such a performance. If anything such a reorganization might give a little more interest to the quartet. The *Presto* following right on the heels of the lyrical *Andante* might perhaps be more stimulating. Nonetheless given the facts of the situation regardless of the musical merit of the reorganization one can say that a performance unwittingly in accordance with such a reorganization would be an incorrect performance of the work. (Evidently conservative attitudes are here at play.) One aspect of a performance that is usually attributable to the piece of music is that of order yet there is no reason why one can't treat even that aspect somewhat abstractly more so than is customary.

Milhaud writes:

> Still haunted by my memories of Brazil, I assembled a few popular melodies, tangos, maxixes, sambas, and even a Portugese fado, and transcribed them with a rondo-like theme recurring between each two of them. I called this fantasia *Le Boeuf sur le toit,* the title of a Brazilian popular song.[1]

The order of *The bull* is remarkable. The recurring rondo-like theme does indeed recur twelve times after the opening statement. The theme is interspersed between melodies in a rigorous scheme of modulations.

> Milhaud's main *reprise* recurs in cycles of four, each appearance a minor third higher in key; after each fourth appearance a special theme, recurring, takes the music downwards a whole step, to begin the next cycle of keys.
>
> Three complete (and continuous) cycles are heard in this fashion and part of a fourth. The fourth cycle, which would have carried the music through sixteen keys (4 × 4) is broken in the middle when the ever-repeated theme finally gets back to its original key of C major.[2]

To understand the structure of *The cow* in contrast with that of *The bull* it is necessary to attend to the four cycles of the latter here designated as *I II III* and *IV*. *I* is the cycle in which the theme is stated in *C* and then modulates in minor thirds to E^b then G^b and then *A*. Cycle *II* begins in *G* and modulates to B^b then D^b then *E*. And so on. *The cow* however has a more abstract structure than *The bull.* Indeed its structure is such that it is not capable of complete physical realization in any single performance. *The cow* is constituted of Cycles *I II III* and *IV* but in random order. Thus for any particular performance the actual sequence of parts is determined by a random procedure.

No. What do you mean no? I mean no *The cow* isn't a piece of music isn't a work isn't music isn't art. Why not? It damn well is no matter what you say. But if you hear it one night and you hear *IV II I III* and then you hear it another night and you hear *II I III IV* are you going to say you heard the same thing each time? Yes yes. You heard the same piece of music the same work the same music? Yes indeed the same thing each time. The piece of music performed can be treated as abstractly as what is said in uttering an utterance.[3]

Then which are we hearing *The bull* or *The cow?* I mean suppose the sequence chosen at random was *I II III* and *IV* that would mean that we might be hearing *The cow* and not *The bull* even though in such a case the two would be acoustically identical. Yes does that alarm you? As art is created as new interests appear new and extraordinary demands may be made on the associated aesthetic community. Appreciation of *The cow* calls for an appreciation of the conceptual background of the performance. There's nothing strange in that. Someone who has never seen a cow and doesn't even know what a cow is is in no position to appreciate another cow Dubuffet's *with a subtile nose*. Fixing an unblinking eye on a painting need not be enough to appreciate the picture. And for that matter an attentive but unreflective ear will not suffice for an appreciation of Milhaud's *The bull.* It is not simply a *potpourri* of pleasant melodies maxixes sambas tangos and the like. Medleys are easy to come by easy to provide. But they are not likely to wear well their main charm deriving from their evocation of the original works there combined. In contrast the appeal of *The bull* transcends far surpasses that of its elements considered in isolation. For one is captivated by the elements in that harmonic and melodic structure.

Despite all this it seems absurd to say that *The cow* and *The bull* are two different works which on occasion have acoustically identical performances. That is absurd and it is not what I am saying though I admit I may seem to be. *The bull* and *The cow* are not two works. There is only one work in question and that is *The bull on the roof (Le boeuf sur le toit)*. Milhaud composed *The bull*. In the interests of philosophico-aesthetic research I created *The cow* a separate work. *The cow* is simply a version of is simply a variant of *The bull*. We are here concerned with only one work *The bull* and a variant of it *The cow*.

It would be a sort of miracle if one composer just happened to compose a work that differed only trivially from that of another composer. Assuming that there was no contact between them and so forth. Given the temporal limitations of the human race thresholds of human hearing and so forth the number of conceivable auditory patterns is at most finite. But even if one were to restrict one's attention to piano works say of fifteen minutes in length the number of conceivable compositions though finite would be prodigious. There is really no likelihood whatever of one composer just happening to write a work that is virtually identical with that of another composer. And because that sort of thing doesn't happen our conception of identity with respect to pieces of music isn't adjusted to cope with such cases. And what would one say if a composer say in Buffalo and another composer in Rochester without any contact between them each produced a score that was absolutely identical with that of the other? Have they composed two pieces of music or only one? Have they created two works or has each created the same work? Or have they each produced a variant of one and the same work? But then who wrote the work?

The striking fact about the identity of pieces of music is I think this:

a piece of music can have a variant but there is no such thing as a variant of a person or a stone or a pea in a pod. (At least not yet.) If you have a person and you have another person who's sort of like him he's a different person no matter how much like him. He's a different person. You have two persons. And the same goes for stones and for peas in a pod. No matter how much alike they are if they are two then there are two of them distinct and individual. But if you have a piece of music and this other piece of music is almost exactly like it not exactly but almost exactly then it's a variant of the first or the first is a variant of it or the two are variants of something else. No matter: what you don't have is two different works. That's not the way it is with things like paintings or stones or people but that is the way it is with poems and books and essays and pieces of music. That's a central fact about identity with respect to pieces of music.[4]

III

A FINE FOREHAND

"As long as a branch of science offers an abundance of problems, so long is it alive: a lack of problems foreshadows extinction or the cessation of independent development."[1] On this occasion here and on his birthday and whenever one is doing philosophy these words of David Hilbert seem particularly appropriate.[2] The aesthetics of sport is supposedly the subject. It is not a serious subject if it has no significant problems. There are no significant problems that can sensibly be characterized as problems in the aesthetics of sport.

Metatheoretic discussions of what is or is not in the province of aesthetics are largely exercises in futility tiresome and fruitless. But conceivably here and now this dismal generalization should be disregarded. Certain research should be encouraged other discouraged. Research devoted to the aesthetics of sport can accomplish nothing. There is nothing there to be accomplished. Worse it would not only contribute to the vaunted dreariness of aesthetics it could serve to delay even impede other possibly significant research. There are philosophically interesting and perplexing features of athletic behavior. There are deep philosophic questions about physical activity that are worth answering. When there are important paths to explore why maunder about in the vacuity of an aesthetics of sport?

In philosophizing about sport one could but need not begin by worrying about what is and what is not a sport. One could attempt to fix limits draw boundaries. I shall not. That's dull matter best left to linguists and lexicographers. It's not as though this were a novel problem of analysis. Drawing boundaries and fixing conceptual limits is generally difficult in nonformal or nonrigorous domains and almost invariably unproductive. Anyway examples of sports are easy enough to come by. Archery is a sport so is auto racing badminton baseball basketball bicycling bobsledding bowling boxing so is bull fighting and canoeing. Cave exploration is sometimes accounted a sport. Curling fencing field hockey fishing football gliding golf gymnastics handball. There's no problem about examples there are all kinds of examples. Judo karate lacrosse are all sports so is pigeon racing and polo so is shooting and tiddlywinks.

If one is concerned with the aesthetics of a certain class of things then the members of that class must characteristically have certain aesthetic aspects. The aesthetics of that class would then concern itself with the aesthetic aspects of the members of the class. If it should prove to be the case that the members did not characteristically have aesthetic aspects there would be nothing for the aesthetics of the class to be concerned with. The aesthetics of sport is in almost but not quite that position. Some sports happen to have aesthetic aspects. Most sports do not. And no novel and significant problem is posed by those sports that do happen to have aesthetic aspects.

Anything has an aesthetic aspect so one might think but is that true? I think not. "The record number of clay birds shot in an hour is 1,308 by Joseph Nother (formerly Wheater) (born 1918) of Kingston-upon-Hull, Yorkshire, England, at Bedford on September 21, 1957. Using 5 guns and 7 loaders, he shot 1,000 in 42 minutes 22.5 seconds."[3] Is there an aesthetic aspect to clay pigeon shooting?

Suppose we suppose ourselves present on that historic occasion to witness Joseph Nother's (formerly Wheater's) achievement. The occasion would have not been an occasion for any form of auditory satisfaction. On the contrary we should have taken care to block our ears to prevent damage. Visually the scene must have been tedious: an hour of seeing clay bird after clay bird after clay bird being hurled into the air by a device immediately to be blasted by the attentive Nother. Even so it could have been (given the appropriate interests) an exciting event to witness particularly if we kept count of the number of clay birds blasted and if we had realized that a record breaking was possibly in the offing. The tension would have or could have mounted as clay birds and minutes moved.

So there is an aesthetic aspect to clay pigeon shooting. Is there? Doesn't even clay pigeon shooting have its dramatic moment? Does it? And does that matter? The breaking of a record is perhaps a dramatic matter but that doesn't mean that clay pigeon shooting is a dramatic matter even if the record in question is a record of clay pigeon shooting.

Perhaps another example will serve us better. Live birds may be livelier than clay pigeons. "The greatest recorded lifetime bag is 556,000 birds, including 241,000 pheasants, by the 2nd Marquess of Ripon (1867–1923) of England. He himself dropped dead on a grouse moor after his 52nd bird on the morning of September 22, 1923."[4] Suppose we suppose that we are with the 2nd Marquess of Ripon on some occasion of this slaughter. An alarmed pheasant breaks cover only to be zapped by the ready Ripon. That could be an exciting moment. And still another example perhaps choicer. Biggest bag: "The largest animal ever shot by any big game hunter was a bull African elephant (*Loxodonta africana*) shot by J. J. Fenykovi (Hungary), 48 miles north-northwest of Macusso, Angola, on November 13, 1955. It required 16 heavy caliber bullets from a 0.416 Rigby and weighed an estimated 24,000 lbs., standing 13 feet 2 inches at the shoulder."[5] Wasn't that an exciting moment?

Was that an aesthetic event? Those present must have been stirred aroused. Perhaps infuriated perhaps nauseated by the mindless sportsman intent on his prey. That the event could excite does not establish the existence of an aesthetic aspect. Not everything exciting is aesthetic. The murder of such a massive creature would be likely to occasion some sort of strong reaction on the part of the spectators. Possibly nausea. I am not objecting to nausea as an aesthetic reaction. But their nausea if that is what it was would have been occasioned by what is not an aesthetic matter.

What if *per impossible* in fact the elephant was not shot was unharmed that all was a sham and a show: Would the event so understood and witnessed still be exciting? I think not. Would it evoke nausea? Not to the same extent if at all and if so for other reasons. For reality matters in sports in a way that it does not matter in art. To blast a clay pigeon out of the air may be minimally exciting but only if in fact one actually blasts it only if in fact there really is a clay pigeon and one really does aim fire and accomplish one's and its end. If such an affair were rigged if say the whole event were subject to the control of a computer and one simply was to go through the motions who would do it? Who would engage in such a practice? But that is not the way it is in matters pertaining to art. To be told that the man acting Lear on the stage is not in fact shedding genuine tears not in fact the least bit unhappy is in fact having the time of his life giving a great performance all that is irrelevant. The excitement one feels in witnessing the play and the sympathy one feels for Lear remain the same.

What I am saying here has been said before. "Now, where the question is whether something is beautiful, we do not want to know, whether we, or any one else, are, or even could be, concerned in the real existence of the thing, but rather what estimate we form of it on mere contemplation (intuition or reflection)."[6] And there is no need to feel mystified by Kant's correct though unoriginal claim. For aesthetic reactions occur in the course of specific acts performed in connection with entities of an appropriate sort. One such familiar act is that of contemplation: the performance of that act in connection with some event does not require that the event be real rather than simulated acted a matter of pretense. There is nothing new here.

Clay pigeon shooting has no aesthetic aspect. The sport is of no aesthetic interest. This is not to deny that one can be interested in clay pigeon shooting. But not every interest is an aesthetic interest. (Perhaps the most interesting thing about clay pigeon shooting is that like all too many sports it is evidently a manifestation of aggression. When one looks at sports in general aggressive not aesthetic aspects are what loom large. Archery boxing bull fighting fencing football judo karate lacrosse shooting wrestling: all offer unmistakable examples of aggressive behavior. As anthropologists and sociologists are beginning to tell us aggressive behavior is most likely learned the result of cultural indoctrination. Our society would be better off without such sports. However there doesn't seem to be any way of putting an end to them not yet anyway.) The sport of clay pigeon shooting has no aesthetic aspect. The same is true of tiddlywinks of shuffleboard of archery baseball basketball bicycling bowling canoeing curling golf fishing.

Doesn't fishing have an aesthetic aspect? Think of casting in a trout stream crystal water aromatic pines surrounding mountains and so forth. All very nice all irrelevant. Does brushing one's teeth have an aesthetic aspect? One could do it in a stream of crystal water in a piney forest and so forth. And one could also fish or brush one's teeth surrounded by garbage smelling smog. Something is not an aspect of an activity unless it serves to individuate that activity. Being in deep waters is an aspect of deep sea fishing: one wouldn't be deep sea fishing if one weren't in deep waters. But standing in a stream of crystal water in a piney forest is not an aspect of fishing or even of trout fishing for one could be fishing or trout fishing even if one were not there but were surrounded by garbage and knee deep in sludge.

Unlike tiddlywinks shooting shuffleboard or archery baseball basketball bicycling bowling canoeing curling fishing golf some sports have distinct aesthetic aspects. This is true of gymnastics ski-jumping figure skating high-diving and even bull fighting. The relevant difference between the first and second group is this: form is a grading factor only for the second. How one does it counts in the second group of sports but not in the first. Sink the ball hit the target: that's what counts in the first group. Form doesn't. Hold the club any way one likes look like a duffer: if one manages somehow to sink the ball expeditiously enough one may end up a champion.

But even though gymnastics has an aesthetic aspect aesthetic factors have at best an inconsequential ancillary role to play in the sport. It is sometimes supposed that there is a difficulty in drawing a sharp line between a performance of a ballet and a gymnastic event. There are similarities between the two. The differences remain obvious unmistakable. Various aesthetically relevant and significant aspects of ballet have no counterpart in gymnastics. A ballet is often a drama: the story is then an integral part of the event. Even when a ballet is without a story it calls for costumes props stage scenery decor that is in accord with the music: there is an intimate relation between auditory and supplementary visual features. None of this is true of gymnastics. Brute strength is an aspect of gymnastics but not of ballet. Suppose a ballet were designed to simulate a gymnastics performance. The dancer might be called on to do an L cross on the rings. Most likely he would have to fake it. Which wouldn't be difficult: thin wires invisible to the audience would serve. Would the fact that he faked the cross make any difference to the aesthetic quality of the performance? Not if it didn't look like a fake (and perhaps not even if it did). And what if a real gymnast at an actual meet faked it? If it were discovered he'd be disqualified excommunicated. The L cross on the rings is an impressive gymnastic stunt. It is not aesthetically impressive. It calls for extraordinary strength or an extraordinary physique. It is no problem for any ordinary chimpanzee or any human who approximates a chimpanzee in appearance. A long-armed large-chested legless freak could do an Olympic cross with ease.

Gymnastics does have an aesthetic aspect. Though considerable strength is required for certain stunts the L cross on the rings front and back levers not strength but balance smoothness ease timing are essential ingredients of various routines. A muscular but inept novice could manage a kip on the high bar by converting it essentially to a muscle-up. Rightly done however a kip on the high bar doesn't require much strength. (Which is why the stunt is best practiced with an under-grip which virtually precludes the possibility of converting it to a muscle-up.) A giant on the high bar calls for strength but timing and balance are also required. Properly executed the giant has an aesthetic appeal. And without a doubt a Hecht dismount is a spectacularly beautiful stunt.

The aesthetic appeal of a gymnastic stunt such as the Hecht dismount from the high bar or for that matter of any properly executed gymnastic stunt is a byproduct of other factors. Considerations of mechanical efficiency and strength are fundamental in gymnastics. To execute a reverse giant correctly it is necessary to keep one's arms straight throughout the swing. Why? Because a giant performed with arms bent is less beautiful? That's far from clear. If a gymnast were performing a routine with a sequence of giants to execute one with bent arms might lend some variety to the event an aesthetically desirable variety. But such a performance would be unacceptable to a gymnastics judge. A giant executed with bent arms is not mechanically perfect: it is inefficient and it calls for much greater strength to cope with the centrifugal force at the bottom of the swing.

Although form does not figure in its scoring tennis has at least on occasion considerable aesthetic appeal. A flat forehand drive executed with perfect form from the baseline is a graceful stroke. Even so aesthetic factors have no role to play in tennis: its aesthetic appeal is simply an epiphenomenon. Anyone who plays the game seriously knows this but it is easy to misunderstand the matter.

Any tournament player is familiar with players who rally beautifully hitting the ball with grace and ease but who play badly. And any tournament player is familiar with players who have awkward looking strokes and graceless movements but who by dint of agility and effort manage over and over again to return seeming put-aways and hence are extremely difficult to beat. It is not looks but points that win a tennis match.

Nonetheless generally speaking it is a fact that improving one's appearance making one's movements smoother more graceful is likely to upgrade one's performance. Making sure that the forehand is hit with a long fluid follow through will not only add grace to one's stroke but it will increase one's control while adding considerable pace to the ball. (It also means that one has less time to prepare for the next stroke. It would be no consolation to know that one would look graceful while the ball was being returned for a winner by an opponent who cared less about looks than about points.) Not all good tennis strokes however are particularly graceful. Laver's midcourt topspin forehand is a wristy flicking sort of stroke having none of the aesthetic appeal of his great backhand baseline topspin drive. Yet Laver uses that wristy flicking topspin forehand with great success. The beauty of tennis is simply owing to the by now familiar fact that beauty is often a byproduct of mechanical efficiency. (Possibly at times mechanical efficiency is itself taken as a standard of beauty.) That beauty is a sometimes byproduct of mechanical efficiency has been known at least since the time that people began treating airplane propellers as objects of art.

The aesthetics of sport is a subject without any significant problems. Aesthetic questions posed by a study of gymnastics or tennis have already been posed and are better posed either in connection with traditional art forms such as the classical ballet or in connection with various forms of modern art such as *l'art trouvé*. There are significant philosphical problems to be considered in connection with sports but they are not problems of aesthetics: they are epistemological linguistic and logical in character.

Consider a flat forehand drive hit from the baseline in tennis. To hit such a drive properly there are various things one should do. Move forward to the ball. Bend one's knees. Hit the ball as far in front of the body as is feasible. Hit it on the rise before it reaches the top of the bounce but not so early as to convert the stroke into a half-volley. And so on. To supply such specifications one describes as accurately and as carefully as one can what a fine player actually does in executing the stroke in question. The task of supplying such specifications is primarily descriptive and analytic. (That there is likely in fact to be some sort of idealization involved is no doubt true but not germane to my purposes here and now.) A trained observer a tennis coach attempts to perform this sort of descriptive analytic task.

Suppose a player is told by the coach that he does not bend his knees enough in hitting the forehand. Presumably the player wishing to improve his stroke will attend to what he was told. How does he do that? Since he was told to bend his knees when he hits the ball he does just that. That means he will move his legs in such a way as to make them conform to the description supplied by the coach. How does he know whether or not he has done that? The coach knows by looking at the player. But how does the player himself know?

In recent years some philosophers have toyed with what they call "knowledge without observation" it supposedly being the kind of knowledge one has of one's bodily position of the position of one's limbs and so forth. One reason why sports provide a fertile epistemological field is that one finds that often athletes (or would-be athletes) do not in fact know the position of their own limbs. Many players think they bend a great deal in stroking a ball whereas in fact they hardly bend at all. This is a particularly common illusion with respect to the service motion in tennis.

Suppose a conscientious player uses a mirror to check whether he is bending enough. Then having rehearsed the stroke before the mirror he attempts to execute it on a court. Evidently he must somehow remember what if felt like when he bent in the appropriate way before the mirror. How does he do that? I am not asking how he manages to remember advice. I take it for granted for the time being that people can remember words and symbolic matters. I do not know how they do that but that is not in question here. What I am asking is how one manages to remember positions of the body. And the form of the answer must be or so it seems at present that something in one's mind constitutes a representation of the body position in question. Is that right? I really don't know but it certainly seems so to me. And if that is right then that suggests that his mental representation of his bodily position plays a role similar to that played by the explicit linguistic description to the effect that his knees are bent in the desired way. Are there significant similarities here?

Assuming that the player does have a mental representation of his bodily position what is the correct or a correct or at least a reasonable characterization of that representation? The term one encounters in psychological discussions of related matters is "image". But that that is a plausible characterization is not clear to me. If that is an image then that sort of image seems radically different from what are spoken of as visual images. Is it an image?

Coaches generally do not restrict themselves to descriptive analytic comments. If a player is having problems keeping the ball on the racquet the coach may suggest that he think of it in a certain way: 'Imagine that you are to hit not just one ball but seven of them all in a row'. Such advice is apt to be helpful. How it manages to be that is remarkably difficult to say. Possibly the player forms the appropriate image in his mind. He then somehow adjusts his movements in accordance with this image. How is that done?

I do not have adequate answers to the questions I have been raising. If I did I should not have raised them here. And there are other related questions that want asking and answering. What is actually accomplished by practicing a stroke? Are effective short cuts possible? If not why not? Many tasks can be learned in a very short time: why can't one learn a fine forehand in a short time? Does understanding how an act is performed facilitate the performance of the act? If it does why does it and if it doesn't why doesn't it? These are just a few of the difficult problems that stare back at a thinking being when he turns his eyes on sports. Such problems as these and not aesthetic trivialities are deserving of a philosopher's careful attention.

IV

QUOTE: JUDGEMENTS FROM OUR BRAIN

In 1938 Wittgenstein gave some lectures on aesthetics. Notes were taken by students and these notes have been published.[1] I shall comment on some aspects of the lectures as indicated by the notes.

There is a matter of style. Wittgenstein said:

> How much we are doing is changing the style of thinking and how much I'm doing is changing the style of thinking and how much I'm doing is persuading people to change their style of thinking. (p. 28.)

The style is romantic. He said:

> The subject (Aesthetics) is very big and entirely misunderstood as far as I can see. The use of such a word as 'beautiful' is even more apt to be misunderstood if you look at the linguistic form of sentences in which it occurs than most other words. (p. 1.)

Is the subject of aesthetics "very big and entirely misunderstood"? The words of Unmon are available: A monk asked Unmon "What is Buddha?" Unmon answered "A dung-wiping stick".[2] But the use of the word 'beautiful' has little role to play nowadays.

Wittgenstein was a romantic. The author of the *Tractatus Logico-philosophicus* was a stylistic descendant of Hegel: the lectures on aesthetics evoke images out of Rousseau:

> There are lots of people, well-offish, who have been to good schools, who can afford to travel about and see the Louvre, etc., and who know a lot about and can talk fluently about dozens of painters. There is another person who has seen very few paintings, but who looks intensely at one or two paintings which make a profound impression on him.[3] [[3] Someone who has not travelled much but who makes certain

observations which show that he 'really does appreciate' ... an appreciation which concentrates on one thing and is very deep – so that you would give your last penny for it.] (p. 9.)

Another koan: what would you do with your last penny? And then there is the Tolstoian echo: "but change the picture ever so slightly and you won't want to look at it any more" (p. 36). What picture? Who won't? Compare: a towering stone construction delicately balanced so that if a single stone is moved the whole will come crashing down: a cantilevered pile immune to earthquakes. The latter supplies the more plausible model of (good) paintings. What "slight" change could spoil *Guernica*? Does the slight strabismus undo Da Vinci's *Ginevra de' Benci*? And what's a "slight" change and what's a big one?

Being a romantic in the 1930s science and the scientist were seen as emblems of idolatry. This perspective is displayed in a comment on Jeans:

Jeans has written a book called *The Mysterious Universe* and I loathe it and call it misleading. Take the title. This alone I would call misleading. Cf. Is the thumb-catcher deluded or not. Was Jeans deluded when he said it was mysterious? I might say the title *The Mysterious Universe* includes a kind of idol worship, the idol being Science and the Scientist. (p. 28.)

The reference to the "thumb-catcher" is explained in a footnote:

I have been talking about the game of 'thumb-catching'. What's wrong with that? 'Thumb-catching': holding the right thumb, say, in the left hand, then trying to grasp it with the right hand. The thumb 'mysteriously' disappears before it can be grasped. (p. 27.)

A recent issue of *Science News* reports a mystery:

This is another to add to the menagerie of pulsed signals, but it is an extremely weird one. The bursts rise to maximum in about half a second and take ten seconds to die down. They occur on the average of every 15,718 seconds, but the repetition is not exactly precise. There is a 'phase jitter' of about 500 seconds one way or the other, the longest recorded discrepancy being about 1000 seconds. Such a difference between pulse length and repetition time is unique in pulsed X-ray phenomena.

Normally, pulsed signals are attributed to pulsing or rotating bodies, but the difference in the numbers and the jitter make it hard to imagine what kind of body could produce these. If the source is indeed in the globular cluster in Sagittarius from the direction of which the bursts come, the intensity of a burst is a million times the intensity of all radiation from the sun.[3]

Some astrophysical aspects of our universe are mysterious. What kind of body can be such a weirdly jittery X-ray source? Is that like asking in the "game" of thumb-catching "Where has the thumb gone?"?

Scientific responses are not what Wittgenstein wanted to answer the questions he was concerned with:

Supposing it was found that all our judgements proceeded from our brain. We discovered particular kinds of mechanism in the brain, formulated general laws, etc. One could show that this sequence of notes produces this particular kind of reaction; makes a man smile and say: "Oh, how wonderful."[1] [[1] If you knew the mechanism of molecules there, and then knew the sequence of notes in the music, we could show that .... – R.] (Mechanism for English language, etc.)[2] [[2] That he says it in English and not in French would also be explained by the fact that something is embodied in his brain: we could see the differences. – R.] Suppose this were done, it might enable us to predict what a particular person would like and dislike. We could calculate these things. The question is whether this is the sort of explanation we should like to have when we are puzzled about aesthetic impressions, e.g. there is a puzzle – "Why do these bars give me such a peculiar impression?" Obviously it isn't this, i.e. a calculation, an account of reactions, etc., we want – apart from the obvious impossibility of the thing. (p. 20.)

Is it impossible? Couldn't we discover particular kinds of brain mechanisms formulate general laws so that one could then show that a certain sequence of notes produces a particular kind of reaction makes a man smile and say "Oh how wonderful"? Nothing of the sort is known today. But there's no good reason to suppose that nothing of the sort will ever be discovered. Some effective method of scanning brain structure may be developed something comparable to the scan now possible with grosser features of the body: an X-ray delta scanner can nowadays supply a computer display a picture of the body discriminating between points a centimeter apart and allowing for the different densities of the tissues scanned. Possibly in time some new type of scanner will scan a particular person's brain supply a computerized analysis of the brain structure allow us to make the predictions in question. Wittgenstein spoke of the obvious impossibility of the thing the impossibility of discovering such mechanisms of formulating general laws. He said

> [The] paradigm of the sciences is mechanics. If people imagine a psychology, their ideal is a mechanics of the soul.[1] [[1] I suppose the paradigm of all science is mechanics, e.g. Newtonian mechanics. Psychology: Three laws for the soul. – S.] If we look at what actually corresponds to that, we find that there are physical experiments and there are psychological experiments. There are laws of physics and there are laws – if you wish to be polite – of psychology. But in physics there are almost too many laws; in psychology there are hardly any. So, to talk about a mechanics of the soul is slightly funny. (pp. 28–29.)

The time was 1938 B.C. – before the computer. What is in question is not a "mechanics of the soul" but a precise detailed explicit computerized analysis of human neuro-physiological functions. And if by some amazing combination of circumstances the race manages to survive such an analysis will soon be forthcoming.

Wittgenstein was concerned with what he called "puzzles" about "aesthetic impressions". A puzzle is supposed to be a question an answer to which is hard to come by for some special sort of reason. At any rate it's not just any question an answer to which is not readily available. For example "What's the exact height of Old East dorm on the U.N.C. campus?" is a question I can't answer at the moment by I don't think that question is supposed to express a "puzzlement". Would that question serve to express a "puzzlement" for some primitive innocent of geometry and without access to records? Wittgenstein didn't discuss the matter.

Wittgenstein was concerned with puzzles about "aesthetic impressions". "There is" he said "a puzzle, why do these bars give me such a peculiar impression" (p. 20). The puzzle comes I shall suppose in various forms. So one might ask: why do these lines (of a poem) give that man such a peculiar impression? Or: why does this face in the painting give her such a peculiar impression? And so forth. The question the "puzzle" "Why do these bars give me such a peculiar impression?" would seem to be a question about the effect of a work or of a part of a work on a person. And that is what Wittgenstein said: "Aesthetic puzzles – puzzles about the effects the arts have on us" (p. 28).

Consider a case: the line "A garden is a lovesome thing, God wot" gives George a peculiar impression: a feeling of delight of escape and release. The puzzle is: why does this line give him such a peculiar impression? It is discovered that George has been the victim of a psychological experiment: he has been subjected to an operant conditioning schedule supplied by ardent Skinnerians. The line "A garden is a lovesome thing, God wot" had been associated with positive reinforcement throughout the period of conditioning: it served to signalize release from unpleasant stimuli. Would this serve to explain why the line gave George this peculiar impression? Wittgenstein said:

> One of the curious things about psychological experiments is that they have to be made on a number of subjects. It is the agreements of Smith, Jones and Robinson which allows you to give a explanation – in this sense of explanation, e.g. you can try out a piece of music in a psychological laboratory and get the result that the music acts in such and such a way under such and such a drug. This is not what one means or what one is driving at by an investigation into aesthetics. (p. 21.)

But if we found that George had been subjected to this sort of operant conditioning wouldn't that serve to explain why the line "A garden is a lovesome thing, God wot" made such a peculiar impression on George? But aesthetic impressions are what are in question. George's feeling of delight of escape and release was occasioned by his hearing the line "A garden is a lovesome thing, God wot" and for altogether explicable reasons. But it is not the sort of feeling or impression one is concerned with (that Wittgenstein was concerned with) in an investigation of aesthetics. Because it's not an aesthetic impression created by the line. But since to say this is not likely to make things much clearer let's turn to another case.

Some paintings make some people feel dizzy: a work of op art can have such an effect. Suppose then someone asks what it is about such a painting that makes him dizzy. The answer might be: the blurred edges blurred in the way that op artists know how to blur edges to make people dizzy. Wittgenstein said:

> The sort of explanation one is looking for when one is puzzled by an aesthetic impression is not a causal explanation, not one corroborated by experience or by statistics about how people react. (p. 21.)

That people are made dizzy by certain works of op art seems to be a fact and op artists seem to know what there is about a work of op art that makes a person dizzy. The explanation that an op artist could supply could be corroborated by experience or by statistics about how people react. The question "Why does this work of op art make me dizzy?" seems to admit of a straightforward answer of the type that Wittgenstein rejected.

Looking at a work of op art someone says "It makes me dizzy" and it does: is dizziness an aesthetically relevant reaction on viewing a work of art? A work of op art yes of course depending on the work. Maybe that isn't the way it was in 1938 but that's the way it is today. Art changes. Nausea boredom indifference any reaction whatever can be aesthetically relevant. If one can lend oneself to the catharsis of tragedy what could require one to avoid the purgation of nausea or boredom or indifference? Nausea can be difficult to account for. Some years ago Meret Oppenheim exhibited a tea cup and saucer covered with rat's fur or mouse fur. The work evoked nausea. Why? I don't know. The matter is I think a question for a psychologist. Did Wittgenstein deny this? I think so.

To stand back for a moment to survey: his conception of art is what is antiquated. "What belongs to a language game" he said "is a whole culture" (p. 8). But his culture is alien to me: "What does a person who knows a good suit say when trying on a suit at the tailor's?" (p. 5). What does he say? "Who can beat Connors in Vegas?" Or:

If you say he appreciates it [Negro Art], I don't yet know what this means. He may fill his room with objects of Negro Art. Does he just say: "Ah!"? Or does he do what the best Negro musicians do? (p. 9.)

What do the best "Negro" musicians do? Sniff horse or just smoke pot? And there is this other "person who has seen very few paintings, but who looks intensely at one or two paintings which make a profound impression on him": I don't want the peasant peering over my shoulder looking at my paintings.

Suppose one asks why one of Graham Sutherland's paintings of a thorn creates the peculiar impression it does. What impression does it create? Looking at the painting one says "It has a feeling of closeness of oppression of a brooding presence." One can say that this Sutherland painting creates this feeling has this feeling. But this is not to say that in looking at this Sutherland painting one has such a feeling. To say that a work creates a certain feeling is not to say that in looking at the work an observer has the feeling created. In this respect the feeling of closeness created by the Sutherland painting is in sharp contrast with the feeling of dizziness created by some work of op art. One could express this difference by saying that the Sutherland work "has a feeling of closeness" whereas the work of op art "makes one feel dizzy".

But (if I read him right) according to Wittgenstein the difference between these two cases is not as great as this way of expressing it can make it seem. For though in looking at the work of op art one may feel dizzy whereas in looking at the Sutherland one is not likely to feel oppressed in each case one does have a certain reaction to the work though a different reaction and one expresses these reactions by saying in one case "It makes me feel dizzy" and in the other "It has a feeling of closeness of oppression of a brooding presence."

On viewing the Sutherland work one may say "It has a feeling of closeness" etc. What impression is in question? Wittgenstein said:

...the audience easily distinguishes between the face of the actor and the face of

Lloyd George. All have learnt the use of ' = '. And suddenly they use it in a peculiar way. They say: "This is Lloyd George," although in another sense there is no similarity. An equality which we could call the 'equality of expression'. We have learnt the use of 'the same'. Suddenly we automatically use 'the same' when there is not similarity of length, weight or anything of the sort. ...The most exact description of my feelings here would be that I say: "Oh, that's Lloyd George!"[1] [[1] Important thing is I say: 'Yes, this is Drury.' If you wish to describe feelings, the best way is to describe reactions. Saying 'This is Drury' is the most exact description of feelings I can give at all. Idea that most exact way of describing is by feelings in the stomach. –S.] Suppose the most exact description of a feeling is "stomach-ache". But why isn't the most important description of feeling that you say: "Oh, this is the same as that!"? (pp. 32–33.)

The aesthetic impression Wittgenstein would have been concerned with here I suppose is that "impression" expressed in uttering the words "It has a feeling of closeness" etc. in reaction to and on viewing the Sutherland painting of a thorn. The "puzzle" for him would have been why this work gives the viewer this peculiar "impression".

Wittgenstein's use (or the use attributed to him in the lecture notes) of the word 'feeling' was odd. If it dawns on me that an actor is playing the role of Lloyd George and I say "Oh that's Lloyd George" I am not expressing my feelings nor if you reported my utterance would you be describing them. But given this odd use it appears that an odd question would on Wittgenstein's view then occasion the "puzzle" about the "impression" someone might receive from the Sutherland work. For the "puzzle" would be to explain why someone reacts to the work by uttering the utterance "It has a feeling of closeness" etc.

Wittgenstein said:

...we can dream of predicting the reactions of human beings, say to works of art. If we imagine the dream realized, we'd not thereby have solved what we feel to be aesthetic puzzlements, although we may be able to predict that a certain line of poetry will, on a certain person, act in such and such a way. What we really want, to solve aesthetic puzzlements, is certain comparisons – grouping together of certain cases. (p. 29.)

Wittgenstein said the sort of explanation one is looking for is not a causal explanation. And he said this is connected with the difference between cause and motive:

> In a lawcourt you are asked the motive of your action and you are supposed to know it. Unless you lie you are supposed to be able to tell the motive of your action. You are not supposed to know the laws by which your body and mind are governed. (p. 21.)

Is that like saying that although someone may not know why the rat cup and saucer makes him nauseous he is supposed to know why he reacts to the Sutherland work by saying "It has a feeling of closeness" etc.? Wittgenstein is then surely right in suggesting that a causal answer is not wanted in answer to such a question as "Why does the Sutherland work give me that peculiar impression?" if the impression in question is taken to be the uttering of the utterance "It has a feeling of closeness" etc. on viewing the work. To give a causal answer to such a question one would have to know what causes him to utter precisely that English utterance in the given situation: but a mechanics of English is likely to be of little interest to one concerned with the aesthetic aspects of the Sutherland painting.

Why then does the Sutherland painting give one the peculiar impression it does? If that is to ask why one reacts to the work by uttering the utterance "It has a feeling of closeness" etc. the answer seems simple enough: someone may say "It has a feeling of closeness" etc. because it does and perhaps because she thinks someone else might like to know or perhaps just because she likes to say things out loud. Then why does the Sutherland painting have that feeling of closeness?

"What we really want" said Wittgenstein "to solve aesthetic puzzlements, is certain comparisons – grouping together of certain cases" (p. 29). There is a footnote to this remark:

> A picture, 'Creation of Adam' by Michelangelo, comes to mind. I have a queer idea which could be expressed by: 'There is a tremendous *philosophy* behind this picture.' –S. (p. 29.)

Elsewhere Wittgenstein remarked:

> As far as one can see the puzzlement I am talking about can be cured only by peculiar kinds of comparisons, e.g. by an arrangement of certain musical figures, comparing their effect on us.[3] [[3] When the written notes or the played notes are spread out, then you say ...–T]. "If we put in this chord it does not have that effect; if we put in this chord it does." (p. 20.)

Why does the Sutherland painting of a thorn have that feeling of closeness of oppression of a brooding presence? There's no need to look for different versions of the work: comparisons would be otiose. An answer to our "puzzle" is readily to be found in the technical features of the work. That the work in question (which is to be found in the permanent collection of the New York Museum of Modern Art) has the feeling it has is directly attributable to various factors. First the section of the painting representing sky is a cadmium orange. Cadmium orange being a strongly advancing color deprives the painting of any airy or open spatial quality. Secondly the purplish greys of the areas in front of the cadmium orange areas are forced into a forward position by virtue of placement and general features of the drawing and this despite the tendency of such colors to recede. The upshot is a general feeling of closeness of oppression. Of course if it were not for other features of the painting e.g. the movement of the main mass and so forth the feeling in question perhaps would not have been created. But given the presence of the other features the salient feature accounting for the oppressiveness of the work is the cadmium orange sky.

Wittgenstein had much to say about aesthetic "puzzles" about the effects the arts have on us. But what he had to say doesn't fit any of the cases we have considered: not that of George's reaction to the line "A garden is a lovesome thing, God wot" not one's reaction to a work of op art to the rat cup and saucer and not one's finding the Sutherland work to have the feeling it has. I think the reason for this lack of fit is that Wittgenstein was concerned with radically different kinds of "puzzles". He said:

> You can sometimes find the similiarity between the style of a musician and the style of a poet who lived at the same time, or a painter. Take Brahms and Keller. I often found that certain themes of Brahms were extremely Kellerian. This was extraordinarily striking. (p. 32.)

And then there is his reference to Michelangelo's *Creation of Adam*: "I have a queer idea which could be expressed by: 'There is a tremendous *philosophy* behind this picture'." The "puzzle" is: why did certain themes of Brahms strike him as Kellerian? And the "puzzle" is: why did Michelangelo's *Creation of Adam* give him the feeling that there is a tremendous philosophy behind the work? Are these in fact the sort of "puzzles" Wittgenstein was concerned with? I don't really know. How can one know? He didn't cite any real cases. But they seem to be. And how are these different kinds of "puzzles" than the ones we have been considering? Only in being vague and indefinite.

Why did certain themes of Brahms strike him as Kellerian? Which themes of Brahms? When did they give him that impression? Did he read Keller before he heard the themes or after? Which works of Keller struck him as Kellerian? All equally or some more than others? Not knowing the answer to any of these questions I have not the slightest idea why certain themes of Brahms struck him as Kellerian. I do not feel mystified by the matter. Or do vagueness and indefiniteness suffice to give rise to a "puzzle"? Wittgenstein said:

> Here you actually have a case different from that of faces. With faces you can generally soon find something which makes you say: "Yes that's what made them so similar." Whereas I couldn't say now what it is that made Brahms similar to Keller. Nevertheless, I find that utterance of mine interesting. (p. 32.)

That he could not say why certain themes of Brahms struck him as Kellerian may be owing simply to the fact that he failed to investigate the matter carefully. Or at all. Faces are easier to compare than words with music. And why did Michelangelo's *Creation of Adam* give him the feeling that there is a tremendous philosophy behind the work? Probably the answer has something to do with Wittgenstein's own view of philosophy what he took to be the subject matter of the painting his attitude towards matters of creation his feelings about Darwin and so forth. Most likely the solution to these "puzzles" would be of no aesthetic interest.

To ask why a work gives a certain impression to a certain person is to be concerned with the specific relation that obtains between the work and the person. But this relation may be of no aesthetic interest despite the fact that the work in question may be a work of art. Suppose attending to a certain work gives someone a certain impression. Why does it do that? Maybe primarily because of him: as in the case of George cited previously. In which case the impression can hardly be said to be an aesthetic impression created by the work. In contrast the impression may be primarily owing to the work in that it would give that impression to any reasonably sensitive discerning sort: as in the case of the Sutherland thorn.

Are there no "puzzles" of the kind Wittgenstein was concerned with that are of genuine aesthetic significance? There are questions without answers that are likely to remain without answers for some time. Looking at an unsigned drawing I say "That strikes me as a Klee." Why does it give me that impression? I say "Look at the lines!" But what about the lines? I can't say. What is in question is a visual analogy. Von Neumann concerned with such matters said:

> About one fifth of the brain is a visual brain, which, as far as we know, does nothing except make decisions about visual analogies. So, using the figures we have, which are not very good, but which are probably all right for an orientation, we conclude that apparently a network of about 2 billion relays does nothing but determine how to organize a visual picture. It is absolutely not clear a priori that there is a simpler description of what constitutes a visual analogy than a description of the visual brain.[4]

V

ABOUT THE APPRECIATION OF THE DANCE

I am going to talk about problems about the appreciation of dance. These problems are problems I think for one who is concerned to view a dance as a work of art and to appreciate what is being viewed. I don't think they are insurmountable problems but I do think they pose genuine difficulties in the appreciation and evaluation of dance and I think these difficulties are then difficulties in forming aesthetic judgments about dance and difficulties in forming any kind of sensible aesthetic evaluation of the whole process. The first thing I think I want to say is that though many people are inclined to say that dance is a language that's really an unfortunate way of thinking. If you think of dance as a language then you may be inclined to approach it to attempt to appreciate it in the way one appreciates a language to attempt to understand it in the way that one understands a language and that just cannot be. It is not a language. I will go on to suggest to you that it may well be thought of I don't know how strictly but it can certainly sensibly be modeled in terms of a symbolic system. But not every symbolic system is a language. I can give you a simple example. All of you can use road maps I trust in one way or another. In order to use a road map you have to read a symbolic system: red lines black lines distances and the like. Again you can understand the use of code flags and the like. In all these cases one is dealing with symbolic systems but not with a language. Now I'll have to be a bit more technical and I'll use some technical terms to explain more precisely to you why dance is not a language and what kind of symbolic system it can be thought of how it can be modeled in terms of symbolic systems. But don't be worried by the technical terminology because I'll explain it. I do like to put things precisely and then to slop around a bit afterwards. So long as I have said it precisely I don't mind the sloppiness.

Speaking precisely then the reason dance cannot be a language is that it lacks the appropriate syntactic and semantic structures requisite for a language. It has syntactic and semantic structures adequate for a symbolic system but not for a language. More precisely still dance lacks in the most general form either metatheoretic or concatenative or recursive structures. Now let me explain that.

Metatheoretic structures are exemplified in the ordinary use of language in a very simple way. We use words to talk about words. I can say the word 'dog' is spelled with a 'd'. In so doing I am using the word 'dog' to talk about the word 'dog'. That's something that's quite characteristic of language. It occurs in some communication systems for example crows will caw about cawing so they have a minimal metatheoretic structure there but they do it to a very limited extent. A papa crow will teach a baby crow to caw in the right way. So in that sense they have something of it.

The second aspect of language that's essential for linguistic structure is concatenation. We join words to form phrases compounds sentences discourses. You can talk of something being red: you can talk of an apple: you can talk of a red apple. Or a more peculiar structure: you can talk of something that's oil: you can talk of a lamp: you can talk of an oil lamp. A red apple is an apple that's red. An oil lamp isn't a lamp that's oil: it's one that uses oil. It's a different structure.

Concatenation doesn't really occur in any interesting way apart from genuine linguistic structures but there are some examples of concatenation in dance. Movements can be joined to movements to form sequences of movements so you have at that level something of the aspect of concatenation. You find the same thing by the way in bird calls. Herring gulls have an attention call and an alarm call and they put the two together and you have a compound call which functions as a compound call.

The third essential aspect of a language is what's called recursive structure. Indeed this is the most significant part of linguistic development apart from concatenation. Recursive structure can be explained fairly simply. In technical terms a function is recursive if it takes as an argument to the function its own value. Putting it in simple terms if I say something like 'It is raining' I can then take that sentence make it into a subject of a sentence and say something about it. I can say 'That it is raining is true.' And then I can take that sentence and make it into a subject and say 'That that it's raining is true is absurd.' What I have said is in fact quite correct. It is not raining. What I have done is use a recursive linguistic structure to produce this complex utterance.

Again you find something like that in connection with animals. In bird song you find recursive structuring of a motif: the chaffinch will sing a motif which is embedded then in a more elaborate motif which is then embedded in a more elaborate motif. And in dance you can find something like that. There will be movements and then more complex movements embedding the repetition of the initial movements and perhaps then still more complex movements.

It is at the syntactic level that you find these things in connection with animal communication and in connection with dance. Let me explain to you the difference between the syntactic and semantic level of a language or of a symbolic system because that's quite important in order really to understand what kind of symbolic system dance is. If one speaks of syntactic relations in a language one is talking of relations between words groups of words words to words so for example if I say 'red apple' I use an adjective and a noun. That's a syntactic relation: the adjective operates on the noun. Syntactic relations relate certain formal elements to other formal elements. It's a patterning. In dance dance viewed as a syntactic system it's very clear that it has an elaborate syntactic structure. Let's consider what the elements of dance are from a syntactic point of view. This is leaving out all questions of meaning now. Just the elements that enter into a dance.

The most obvious thing one has of course to begin with would be the dancer. That is to say a moving object. The object is generally a person. It needn't be a person. It could be a robot for that matter but generally customarily it's a moving person. A person who is capable of various movements. Persons groups of persons are the central objects in dancing but of course there are other objects. These are the objects that constitute the props around the person. They may be scenery chairs anything. There need be nothing of that sort but there may be.

Then there are the other objects which serve to define the space in which these movements take place: a cyclorama a stage prop of some kind a backdrop of some kind. They may or may not be these. There will be something that serves to define the space however. So we have essentially two kinds of objects to begin with: one the moving ones that we call the dancers the other the objects which surround the dancers and which serve to define the space in some way. And then the matter is complicated even further. There are two other elements of a radically different sort. Both of them or anyway at least one of them absolutely essential to the dance and that is light. If there is no light and the dancers are dancing in say the Carlsbad Caverns without any light one is not going to have much to appreciate. One may get some auditory phenomena but that's about it. So one has light and the light can of course be enormously varied spotlights floodlights any kind of light and all kinds of patterns. They could be stable or not.

And then in addition to light of course there is sound. That's the other fundamental factor. The sound can be music it can be agitando the sound of the flamenco dancers' heels whatever one likes but there is sound. So we have sound light objects in a defined space well or poorly defined. (I am not for a moment suggesting the space has to be well defined.) These provide the syntactic basis for a dance. Still speaking now at the syntactic level what one has is change a change in these configurations. The dancers will move in one way or another. I toyed with this idea for a while: could you really say you had dance if you simply had immobile persons on stage who did nothing and only the lights moved? You had the lights flashing back and forth which could indeed create the appearance of movement. I think you might say it was a dance of lights but it wouldn't be a dance in the ordinary sense in which we speak of a dance. And if nothing moved not even the lights then I don't think it could be said to be a dance: or if it's a dance it's a limiting case of a dance.

The syntax of dance poses enormous problems. These now are not problems of appreciation but I'll digress for a moment to mention them. What's the relevant difference between different changes of this configuration the configuration of persons dancers the objects the lights and the sounds? Say we have some change. What's the difference between those changes of the configuration we are presented with which constitute the performance of a dance and those which do not? I mean suppose we got a bunch of people up on the stage just sort of staggering around. I don't mean staggering around the way a dancer would stagger. I mean just sort of walking around or shuffling about the way a crowd will shuffle about. Presumably one would say if that's a dance it's a very bad dance but how does one draw the line between what counts as a dance and what doesn't? With respect to that I'm afraid I can say nothing illuminating because I haven't the foggiest idea how to draw that line. I think I could have drawn that line 50 years ago. Then it wouldn't have been too hard because all kinds of movements were ruled out.

One has an analogous problem in connection with works of art in a gallery. What constitutes a work of art? This at one time wasn't very hard to say. One would say a statue or a painting something like that. I recall going into a gallery not too long ago and tripping over a bunch of potatoes on the floor and I said 'What is that?' and somebody said 'You have tripped over my work.' If that's a work of art and I'm not joking I'm not saying it isn't then I don't see how to draw the line between what constitutes dance and what doesn't. I think you do get to a limiting case if you have absolute stillness no change in configuration. But apart from that I don't see how to do it.

You might say there ought to be a way of doing it and I sympathize with the feeling but I don't see how to do it. Anything I think of about the movement like the movement has to be interesting and expressive can always be countered by saying you simply haven't cultivated a taste for this kind of movement. Let me give you an analogy. It used to be thought that if you could say something was boring and evoked nausea then that couldn't be a work of art but many many works are designed to be exactly that. To be boring or to evoke nausea. And some of them are profoundly interesting in the way they do do this. Some of Andy Warhol's movies are profoundly boring. I am thinking of one about a kiss that lasted for I don't know how long.

That's one of the problems that occurs in connection with syntax but that's not a problem of appreciation: that's more a problem of I suppose creation. Where can one draw the creative line? How does one say this is and that isn't a dance? But the syntax of dance does pose an enormous problem for the audience. If we are concerned with moving persons moving objects of any kind a change of configuration which depends upon light and sound and the relation between these moving objects and the backdrop in the defined space then where you see it from is an enormous problem because clearly if you sit on one side of the auditorium viewing this space and I sit on another side we do not see the same thing. It will be quite a different configuration.

You can say that this is also true of a piece of sculpture but the thing about a piece of sculpture is that you can walk around and come back to the original spot. When one views dance ordinarily one has a given position in the theatre and one cannot really walk around the orchestra pit changing seats constantly. Anyway if you did you would be missing a lot of the show. So one is stuck with a very limited perspective. It's extremely difficult to see ballet in fact to see modern dance to see any dance unless you're in a very privileged position you are the only person in a very small space and the dancer is dancing for you. Then you might see just about everything. But with any ordinary presentation in theatre if you are high up you see one thing if you are low down you see another. Lateral movements can become invisible from certain places. Certain patterns which are visible from overhead are not visible from below. A choreographer may be inclined to draw a bunch of scribbly lines for the dancers going across the stage. That the lines are scribbly will only seem so from some perspectives. From others they will take on totally different shapes.

Now you might say that if you have sufficient knowledge of the dance knowledge of the tradition and so forth that regardless of where you are you can take in what the movements are. I have thought about that and I don't believe it. I started to reflect on what the topological problems would be. This would mean visualizing a moving line in space and trying to determine in a three-dimensional space from one position exactly what the trajectory was. Or think of them as vectors forces moving in space which probably would be more accurate: there is no way absolutely no way you can visualize this particularly if the dance pattern is novel unfamiliar. You simply cannot capture all of this. You might say the same is true isn't it of watching any sort of theatre art: you don't see everything.

There is another difficulty of appreciation. Now still at the syntactic level I haven't moved to the semantic problems yet the difficulty is this: with a play you see certain things from certain sides of the stage but you can do something with a play that you cannot do with dance. For one thing you can read the play in advance. You can't do that with a dance there is no way. Secondly you can see it again. Whether or not you are seeing the same play again is fairly easy to determine. I don't see how I can give you any strict criteria. That's another matter. That's rather hard to do. But nonetheless one can be fairly precise about this and say yes this is the same play no doubt not being performed in exactly the same way it's a different sort of rendition but still it's the same play. How does one show this? One takes the script. The script will show it. It's the same thing going on. So you can see *Julius Caesar* in modern dress. It's the same play though a radically different interpretation. So one can identify the play on the basis of the script and identify different interpretations of it.

When one turns to dance there's no way in which one can do precisely that because there is no script like that. There are notes by the choreographer: these may be fairly good there may be a notation of some type that will be pretty good but this doesn't even begin to capture all the other things going on: the lighting the sound and the particulars of the arrangements the particular movements that take place at that time. One thinks yes this is an interpretation of the dance and we know what the dance is. I don't doubt that in some sense we do but I don't really see how we can pin it down nor do I know how we can become very precise about this. So to put the point very simply I don't know how you know that you see the same dance twice whereas I do have a better idea about how you know you see the same play twice and the fundamental difference is that with the play there are fairly explicit stage directions and there is a very explicit script and there isn't anything as explicit as that in connection with dance.

Maybe some day there will be some splendid notation devised which will govern everything and if that were so then this problem could be alleviated. It still would not alleviate the problem of the person trying to see what's happening the problem of which perspective to view it from.

Now I'm not suggesting that this makes it impossible to enjoy the dance you see but I do think this makes it extremely difficult to give any kind of critical evaluation of what you see because it may be that you have chosen a particularly bad spot in which to view the ballet or the dance.

Quite possibly from one perspective it's rather poor and from another it is not. Contrast this by the way with what you do in listening to a piece of music. There is an acoustical character to halls to music halls such that if you have a good one it really doesn't matter where you sit – oh within certain limits it will if you have an abominable seat behind a pillar which echoes or something like that.

But go to something like the Academy of Music in Philadelphia almost all the seats are good seats and you can hear pretty much the same piece of music. The same would not be true for example if you were listening to something in the Church of St. Marks in Venice when say there is a performance of brass choirs by Gabrielli. There it's very important that you sit at exactly a certain place because Gabrielli used the resonance of that Church in a very special way. The brass choirs would echo and the reverberations would come to a certain point and if you are at that point you can hear what he wanted you to hear. If you are not you won't. So you can't possibly take just any seat there. But in most acoustical works it's relatively indifferent where you sit. That's not true not at all true with dance. It is relatively indifterent where you sit in connection with a motion picture. If you sit too close to the screen there will be enormous distortion for a short time but your eyes compensate for this owing perhaps to all kinds of gestalt features of perception.

These are difficulties in forming a critical evaluation of a dance: the difficulty of viewing the same work twice and the difficulty of making sure you had a reasonable perspective on it. Much of what the choreographer did may be invisible to you if you have a bad seat.

I think much more complex difficulties occur at the semantic level of this type of symbolic system. To speak of semantics in connection with any kind of symbolic system is to speak of a connection between the syntactic elements the objects the words or anything that constitute the system and something else which can be said to be that which these entities mean or that which they serve to express and the like. So for example if I say the word 'dog' is used to refer to a four-legged canine I'm telling you something of semantic import about that word. I'm telling you what that word is used to refer to. Or if I told you the word 'perhaps' means the same as the word 'maybe' I'm telling you that they are virtually semantically equivalent. They are not strictly semantically equivalent because although they have exactly the same meaning they don't have exactly the same tone.

The connotation or suggestion of 'maybe' is slightly lower class than 'perhaps'. You can find this out by doing field research in linguistics. So I like to give talks in which I say 'Maybe I'll do this and perhaps I won't.' But that's the only difference between those two. So I'm telling you something about their meaning.

When one turns to a symbolic system like that of dance what one finds is not that the principal elements in all this namely the motions of the dance or if you like the changes in the configuration before you have a more general meaning nor that these particular configurations or motions have any specific meaning. They may on occasion. There are trivial cases where one can get something like a straightforward propositional meaning a statement of some sort. In pantomine one can say 'Yes he is saying that his dog is over in the corner.' Or there may be a kind of story associated with it. There may be a written program with it. You can then associate the written program with the details of the story with the gestures of the dance. But generally there is nothing like that which is not to say that the elements of the dance are devoid of any semantic value. Quite the contrary: they have a much more interesting semantic value than that. On occasion they have that trivially but what they genuinely have is expressive character. The movements of a dancer can be profoundly expressive. I needn't remind you of how expressive movements can be. Take the most trivial thing a threatening gesture. You can see this as threatening. You can feel it as threatening. There are all kinds of movements that are expressive.

I said to you that though dance may have something of a recursive structure from a syntactic point of view that is syntactically speaking the elements are woven around other elements and all of them woven around others: at the semantic level there is no recursive structuring. If you have some gesture which is expressive and another gesture which expresses something quite different you cannot just put the two together and get the result to express what the earlier two expressed individually. You can't say take a word like 'oil' and a word like 'lamp' each of which has meaning put them together and then get the meaning of 'oil lamp' and then you cannot say 'That there is an oil lamp is interesting' and 'That that there is an oil lamp is interesting isn't at all interesting.' You can't make that kind of move.

The meaning of complex movements the significance of complex movements in dance is not a simple function of the meaning of its constituents. It doesn't build up recursively in the way linguistic elements build up.

What kind of meaning what kind of significance do these movements have? How is it to be characterized? If you think of dance as a language which is the thesis I am inveighing against then you would be inclined to say that if these movements are part of a language they then must mean something and if they mean something we should be able to say what they mean and if we can't say what they mean then something is wrong here. Because if they do mean something if it is linguistic why can't you say it? I mean there is no such thing as one natural language which completely fails to communicate what some other language will communicate in one way or another. I don't mean to say that what can be said in one language can exactly be said in another. That is not true. But certainly one can get something of what is said. You can approximate to it in various ways. But there is no way of saying what is expressed by a movement in dance that will express it to the extent or can capture it in the way that it's expressed in the dance.

This makes it sound very mysterious. That's because you tend to think of it as a language. So let me give you some analogies here to try to characterize what kind of expression is involved. I say that movements in dance are remarkably expressive and indeed many dancers many choreographers seem to be primarily concerned with it. I was reading an account by Jose Limon of what he wanted to do in a dance and how he would approach a certain topic and everything he said pointed in this direction: the movements had to be profoundly expressive. He characterized what they were supposed to express in very abstract terms. But he didn't try to say what each movement was supposed to do even though he thought it very important that each movement be terribly expressive.

Let me give you an analogy. If one looks at certain objects like a mountain or a mountain range one can see these mountains as impressive as powerful as awesome as brooding as having all sorts of characteristics. Psychologists particularly Gestalt psychologists here speak of physiognomic characteristics: characteristics of objects which serve to evoke certain feelings certain emotions certain attitudes.

I'll give you a very nice example. Somewhere around Phoenix or Tucson there is a rather famous range of mountains called the Superstition Mountains. When I was down there I wanted to go and see them and I tried to find out where they were. Someone told me just to drive down the road and there will be several mountain ranges and then I would come to the Superstition Mountains. And I said 'How will I know that I'm at the Superstition Mountains? I mean how do I tell these from any other mountain ranges because there is just one range after another? And they said 'Don't worry about it when you get there you'll know it.' So I drove and when I got there I knew it because they were impressive they were awe-inspiring. You could see why these would be mountains that would be sacred to the Indians whereas the other mountain ranges didn't have these characteristics at all.

What did I see? Strictly speaking I saw a certain configuration of rocks and the like and there's nothing that I could say about this configuration of rocks that would really clue you in to what makes them so awe-inspiring. They were very tall but so were lots of the other mountains. They were kind of jagged. I'm inclined to say they looked lonely but when I say that I'm already moving to its expressive character and not telling you anything strictly about its physical form.

Let me give you another example. Most of us can read the expressions on other people's faces particularly in our own culture. If it's not our own culture that's quite another matter but intracultural expressions aren't all that difficult to read. If somebody comes up to me and has a threatening look looks hostile looks angry I know it. Generally I know it not invariably you can be mistaken sometimes. There is a good friend of mine a philosopher who has an angry look. He was born with an angry look. You look at his face and you would swear he's angry. But I know him and I know that's not an angry look. That's just the configuration of his face. But if an ordinary person had that look on his face you would say 'Yes he's got an angry look.'

We can all do this. How do we do it? Never mind that for a moment. Let me give you another example. This is a famous example. I got this from the late John Austin a philosopher at Oxford. It was one of his favorite cases. Someone in England designed a teapot that had a marvelous spout that would in fact not drip. Not a single drop would fall off of that tea spout. It would pour into the cup and leave no residue. But when you looked at the spout you would swear that the thing would drip. It had a drippy look. As a result no one would buy it. They would say that thing has got to drip and you can't pour tea from it. It had a rounded spout I mean a big thick spout and you would just swear that the water would drip all over the place. So no one would buy that teapot. Why? Because they quite rightly said it had a drippy look and if it has a drippy look then most likely it would drip. In this case it wouldn't.

Another case another one of Austin's. He was shown through a prison and he turned around and looked at one fellow and said 'He looks like a criminal.' And someone said to him 'That's the warden.' Whereupon Austin replied 'Well he has a criminal look anyway.' Which is quite right. One can recognize certain looks as having a certain character as expressing certain attitudes as displaying certain attitudes. How is this done? I think unfortunately unfortunately for the sake of dance critics the people who like to write about dance and particularly with respect to the expressive aspects semantic aspects of dance I think most likely the answer is that there is no analysis of this available. That there may well be even from a perfectly solid theoretic point of view no way of analyzing those features of a face which make it an angry-looking face a cheerful-looking face no features of a mountain which will enable you to analyze it as being awesome morbid brooding and so forth. The idea that there must be an analysis of these complex features seems to me simply naive. I'm not saying that you can't mention salient features. If you want to mention a sad face and tell someone how to draw a sad face you say turn the mouth down draw someone's mouth so that it curves down. A happy face draw the lips curving up the way you do clown faces. That's easy to do. But all of us know that we can see a clown with his mouth drawn up looking very sad and there is no way you are going to say that the salient features determine inevitably what's going to happen. You can have any of those features and still not get the expression or you can get the expression without any of those features.

I can't prove this but I can give you a fairly simple case: strictly speaking it's not simple but it seems to be. If you start to draw on a blackboard just with a little chalk – if there were a blackboard here I would draw you some of these – shapes and configurations of dots and configurations of lines and so forth such that you can say you see a triangle then how many different kinds can you form and what would be the analysis of the situation such that you could get a computer to identify each one of these as a triangular shape?

As far as I know there is no known theoretic solution to this. The ways of determining whether something is a triangular shape do not admit of analysis.

The late John von Neumann one of the great mathematicians and essentially the creator of the modern computer discussed this problem once the problem of what he called visual analogies and he pointed out that if we take seriously the devices the physiological devices and mechanisms available to human beings which enable them to identify something as triangular it must be a system as complex as a computer with 10 billion relays because one is using approximately one-fifth of the human brain which is the visual cortex. You have in each one of your eyes approximately 100 million receptors. If you think of the complex network that would be required to simulate this to get a computer to do what a human being can do the computer would today have to be something of the size of Texas: it would be a gigantic machine. We are nowhere near that yet. All these people who talk cheerfully about robots walking down the street next week are just perfectly silly. We may get a robot walking down the street I mean the kind you see in kids' movies but you are not going to get anything that will do what human beings can do not in the area of visual perception. Not for a very long time. I don't mean to say it's impossible. It's just technologically unfeasible at the present and probably not feasible for the next 100 years.

The complexity of the analysis that people are capable of using the visual cortex is really staggering. In consequence we are able to identify things as sad gloomy morbid friendly and so forth. We can feel and react to expressive characteristics and gestures even though we cannot spell out how we do this: we can do this of course in various in very many areas. All of you know perfectly well what a table is or a chair is but I doubt that any of you can define the word 'table' such that it will give a correct specification of all and only those objects that you are prepared to call 'a table'.

You can use lots of words without being able to say what these words mean even though you know what they mean. But just so you can recognize and appreciate the expressive aspects or movements of the dancers' movements without being capable of any analysis. Does this mean that the critic is in a hopeless position? The person who tries to analyze what's going on he's in a hopeless position? Yes and no. If the aim is to analyze the expression in a dance thoroughly completely then you might as well forget about it. Go home. It can't be done.

If the aim is to make clear to others one of the interesting and salient features of the expressive movements performed in the dance then the person who is concerned to criticize the dance from that point of view can be of great help. What one needs is a very very sensitive human being who can perceive these things who is thoroughly familiar with the dance so familiar that they can see what they in fact do not literally see appreciate the character of the movement going on on stage and thus overcome the difficulties posed by the syntactic aspects of this symbolic system.

Someone sensitive to the expressive aspects of the dance can serve to make it clear to others. Such a person can help others notice things they didn't notice and attend to things they didn't attend to. And for that reason I think we need someone we need very good critics very good analyses of what's going on in the dance. All I'm suggesting is that you mustn't suppose that there should be anything like a complete analysis and you mustn't look for or you will just be bemusing yourself you mustn't look for anything like the kind of analysis you might get with an actual language.

This kind of symbolic system most likely will permanently resist analysis. Let me give you one analogy here. It's fairly easy in books on harmony and on contrapuntal technique to identify what's fairly good harmony and what's fairly unusual or bad harmony. It's not too hard when one deals with the vertical component in music the harmonic structure to detail it and give guidelines for it.

It is I think virtually impossible to give any adequate guidelines for the melodic structure of music. If someone asks 'What makes a melody a melody?' I think there is no good answer to that and how do you find out whether something is a melody? There's only one way: you try it on a sensitive ear a sensitive human ear. It's quite possible and there are very good theoretic reasons for thinking this might well be so that there is no analysis possible of what constitutes a melody. A melody may be like what's called a book code something that cannot be cracked in any way unless you have the book and it could be that a melody is a melody because something uniquely structured in the human brain is roused by it. There need be no other explanation.

I think the same may well be true of the expressive movements in ballet and dance generally. There need be no good reason why these movements are expressive save that human beings are innately structured to respond to them. On the other hand there may be some movements such that some analysis is possible particularly if it is not innately determined that they be so expressive. And critics can be of great help there. Persons concerned to analyze the dance may indeed come up with something very useful.

So I suggest that these are the difficulties one has to attend to in trying to appreciate dance and I think for that reason one could say that dance is vastly more interesting than a language. It calls for extreme sensitivity close attention and of course there are all sorts of aspects of the dance that I haven't even alluded to but this alone would make it something worthy of attention because it's difficult and being difficult is something very good in connection with art because you don't quickly get tired of it.

VI

ART AND SOCIOBIOLOGY

Sociobiology is "the systematic study of the biological basis of all forms of social behavior, in all kinds of organisms, including man".[1] The forms of social behavior displayed by man *homo sapiens* are vast in number and fantastically heterogeneous. But from a biological perspective none are more remarkable than those exemplified in sophisticated contemporary aesthetic practices: the practices of art and art appreciation. Can sociobiology contribute to an understanding of contemporary aesthetic practices?

Some say it cannot: for art is a manifestation of culture and "...biology, while it is an absolutely necessary condition for culture, is equally and absolutely insufficient: it is completely unable to specify the cultural properties of human behavior or their variations from one human group to another".[2] That culture constitutes a distinct agency in human affairs cannot be denied. But as the geneticist Dobzhansky has pointed out "it should not be forgotten that this agent [culture] is entirely dependent on the human genotype".[3] Culture with its capital see situated in the field of anthropology appears to be the latest redoubt in the retreat from reason: appeals to supranatural entities having proven inefficacious suprabiological agencies are now invoked to establish man's sovereignty in the biosphere.

Although there are millions of species presently extant on this small planet aesthetic practices are exemplified solely in the behavior of *homo sapiens*. This remarkable fact appears even more remarkable when one considers how many other forms of behavior are not similarly unique to the species: sleeping rising in the morning washing grooming gathering hunting preparing food feeding nursing the young playing using tools communicating capturing domesticating building repairing farming recruiting educating camouflaging – all are exemplified in the behaviors of other species.

From a biological point of view the most striking aspect of some contemporary aesthetic practices is the fact that they are explicitly and avowedly nonutilitarian. This nonutilitarianism is a relatively recent development in the history of art dating back in the Western world perhaps only to the Renaissance development of easel painting. The art of the Palaeolithic and Aurignacian peoples was not art in the modern sense but ritualized magic: it is not part of our aesthetic practice to hurl spears at murals. Neither were the magnificent works of Egyptian art art in the modern sense: we do not commit our works with the dead to the darkness of a tomb: modern paintings are designed to be viewed by the living. Possibly nonutilitarian practices occurred in connection with Greek pottery despite the evident utilitarian provenance of the objects. But it was not until biologically speaking yesterday or even this morning that nonutilitarian aesthetic practices came into full view.

To focus more sharply on the specific form of behavior that is pertinent here consider the acts and actions aesthetically relevant in connection with such a work as Seurat's great masterpiece *Le Dimanche Sur La Grande Jatte:* the work is displayed exhibited in a well lighted gallery with space to view the work at various distances. The work is viewed attended to visually. The particular acts performed by the viewer may be divided into two distinct classes: ancillary and constitutive. Such ancillary acts as identifying the work determining the date of execution ascertaining the character of the pigmentation employed locating its position in the tradition and so forth are all propaedeutic to an aesthetic appreciation of the work. Any fact at all about a work may facilitate or enhance an appreciation of the work: in consequence there are indefinitely and indeterminately many ancillary acts that may be appropriate in connection with a given work. In contrast acts that are genuinely constitutive of an aesthetic appreciation of the work have a distinct logical character and are comparatively restricted in number.

The word 'appreciate' is commonly used to express gratitude of a sort as in the utterance 'I appreciate your efforts on my behalf.' But the primary and aesthetically relevant sense of the word is best seen in connection with logistics. After the first World War Winston Churchill was asked by the British Admiralty to write an appreciation of the Battle of Jutland a naval engagement. He was not asked to express feelings of gratitude. Prior to a military action commanders of the various groups may be convened for an appreciation: to size up the disposition of the enemy forces materiel and so forth. One can in this sense appreciate a position in chess a political situation a work of art. (It is not difficult to see that the common use of 'appreciate' to express gratitude is merely a derivation from the primary sense of the word. To say 'I appreciate your efforts on my behalf' is primarily to say that one has sized up the situation and has the appropriate emotional response. That that response be a feeling of gratitude is only a matter of convention. For one can easily cancel all indications of gratitude as in 'I appreciate your efforts on my behalf and I mean to pay you back' said in a situation in which one has evidently been done a disservice.

To appreciate a work of art such as Seurat's *Grande Jatte* one attends to it visually: one contemplates the work. Thus one performs an act of aspection for contemplation is but one among many relevant acts of aspection.[4] The first logically interesting feature of an act of aesthetic aspection an act constitutive of an aesthetic appreciation of a work or more precisely of an entity of any sort is that it is non-terminating. The act of running for ten minutes for example is a terminating act: after ten minutes of running it is concluded. It reaches its logical termination. In contrast the act of running is nonterminating. Drinking a fine bottle of cognac is again a terminating act. One may prolong it by sipping slowly and infrequently but if one persists it will reach its logical termination when the last drop is imbibed. In contrast a great work of art is an endless cornucopia largely in virtue of the nonterminating character of the relevant act of aspection.

The second logically interesting feature of an act of aesthetic aspection is that it is potentially of intrinsic value: if all relevant factors are in accordance the act is intrinsically valuable in itself as an end. The value of the act is to be found in the performance of the act itself: thus it has a nonutilitarian value.

The third logically interesting feature of an act of aesthetic aspection is that the entity viewed is viewed as a complex. Appreciating is a matter of sizing up: one cannot size up a simple point in space. Thus contemplating one's navel or a lotus or other such meditative or religious practices is not an act of aesthetic aspection.

Consider viewing a full moon in a clear cloudless sky low on an undifferentiated horizon: many may enjoy the sight but do they in so viewing the moon perform acts of aesthetic aspection? It is I think clear that one must recognize a continuum of such acts. The most primitive forms are constituted simply by some attention to an entity with some response to its physiognomic characters. A three year old child may spontaneously respond to the sight of a brightly colored macaw by bursting into tears to the sight of a koala by bursting into laughter. And so one may view mountains sunsets full moons ocean waves and respond to their physiognomic characters. They may be seen as awesome brooding joyous and so forth. I would not doubt that such primitive forms of aesthetic aspection are virtually universal with *homo sapiens*. And of course given such universality a genetic basis for the behavior is at once implicated.

In contrast more sophisticated forms of aesthetic behavior involve maximal attention to the entity together with maximal attention and response both to physiognomic characters and to configurational matters. Thus the impact of Seurat's *Grande Jatte* is attributable primarily to the physiognomic character of the overall configuration. On the other hand the impact of some of Klee's marvelous works is attributable primarily to the physiognomic character of various details. One can talk on and on about Seurat's *Grande Jatte* detailing specifics of its organization design structure spatial relations and so forth but when one turns to a Klee there is often little more to say than 'Look at the incredible expression on the face.' This is not to suggest that the overall configurational aspects of a Klee have no role to play: he was a genius of flawless taste and taste is a configurational matter. Nonetheless the appreciation of Seurat's *Grande Jatte* or of virtually any of Seurat's works calls for a more sophisticated form of aesthetic behavior than the appreciation of some of Klee's. One would accordingly expect such works of Klee to have the greater popular appeal and this is in fact the case.

Aspection of a work of art such as Seurat's *Grande Jatte* is then an instance of the form of behavior under consideration here. It should be reasonably clear that this is a highly specialized form of behavior that not only is not displayed by any species other than *homo sapiens* but is not displayed by vast numbers of humans. For example one needn't turn to primitive impoverished societies such as the Ik of Uganda whose single stated value seems to be food: sunglassed tourists dutifully "doing the Louvre in a day" no doubt perform various acts ancillary to but hardly any constitutive of an aesthetic appreciation of the works they view.

Why don't falcons for example perform even primitive acts of aesthetic aspection? They have remarkable visual acuity and pattern recognition. A falcon can discriminate between the letters of a conventional eyechart at a distance of a hundred yards. The predatory skill of falcons indicates that they are capable of a logistic appreciation of a situation: they circle overhead select their prey and at a precise moment elect to dive at speeds up to 200 miles an hour. Furthermore the practice of falconry testifies to the fact that falcons are quite capable of learning.

Falcons do not perform acts of aesthetic aspection because they cannot and the same is almost certainly true of all organisms on earth other than some primates. For all organisms other than some primates appear to be incapable of self-awareness and self-attention. Only anthropoid apes and men are known to perform behaviors indicative of self-awareness and self-attention. Thus chimpanzees and men can recognize themselves in a mirror and can use a mirror to groom themselves. In contrast it is impossible to get a cat to see itself in a mirror. The first time a cat confronts its own reflection it may react as though another cat were there: it may arch its back ready to fight. A single sniff may then suffice to make it forever ignore the reflected image and this despite the fact that it may continue to use the mirror to track other moving objects in the environment. It is I think altogether plausible to suppose that the capacity for self-awareness is genetically based and that whatever factors gave rise to it occurred in conjunction with or possibly subsequent to the hypertrophy of the primitive head.

That self-awareness is implicated in an act of aesthetic aspection is not difficult to see. Such an act is constitutive of a potentially intrinsically valuable nonterminating act of appreciation. In viewing a work aesthetically one sizes it up and so to speak savors it again and again: one attends to one's viewing of the work. The savoring of food is another bit of behavior peculiar to primates the performance of which also calls for self-awareness: to relish the food one eats one must attend to one's eating: a cougar may eat a deer but only a primate can relish the taste of venison.

Why do only some people perform sophisticated acts of aesthetic aspection? To gain some insight into the matter one must attend to the phenomenon of culture. But there is no need to adopt a view of culture as a mystical suprabiological agency intervening in human affairs. Such a view leads to nothing but an obfuscation of fundamental biological issues.

Cultural phenomena are to begin with remarkably vague and amorphous. Contrast genetic inheritance with so-called "cultural inheritance". Genetic inheritance is acquired at the moment of meiosis in sexually reproducing organisms at the moment of mitosis in others. Man's total genetic inheritance is found in the human zygote. Half of one's genotype is contributed by one's father half by one's mother. What is acquired is perfectly clear logically speaking. But what does a man's cultural inheritance consist of? There is no particular moment of time at which one receives a cultural inheritance: neither is half attributable to each parent. One's culture is a fuzzy set of behaviors (including verbal behaviors) responses objects and a position in an ecological niche.

The formation and maintenance of a culture or what I shall speak of as "culturation" is a biological process akin to speciation. "Species are groups of interbreeding natural populations that are reproductively isolated from other such groups".[5] A factor likely to be of significance in speciation is geographic (or more precisely spatiotemporal) isolation. Geographic isolation is a relative matter dependent on the mobility of the population in question. Geographic speciation occurs when an interbreeding geographically isolated population develops genetically based isolating mechanisms that serve to guarantee reproductive isolation. Thus in geographic speciation geographic factors are replaced by genetic factors in guarding the integrity of the gene pool.

Geographic culturation is the analogue of geographic speciation. Geographic culturation occurs when an interbreeding geographically isolated population develops phenotypic isolating mechanisms behavioral patterns that serve to promote reproductive isolation. The fundamental difference however between speciation and culturation is that speciation directly preserves the integrity of a gene pool whereas culturation directly preserves the integrity of a phenotypic repertoire and only indirectly serves to preserve the integrity of a gene pool. Migrants entering and interbreeding with a population could conceivably become fully acculturated and thus their presence would not effect the integrity of the phenotypic repertoire but the invasion of such a group would inevitably have some effect on the integrity of the gene pool. Furthermore the behavioral patterns operative in speciation such as courtship dances mating rituals displays and so forth are directly focussed on mate selection whereas many of the behavioral patterns operative in culturation have no such immediate focus. Nonetheless it is I think perfectly clear that culturation does operate albeit indirectly to preserve the integrity of a gene pool. Phenotypic traits both morphological and behavioral are an expression of the interaction between genotypic and environmental factors (including the biotic environment). But populations on different sides of cultural boundaries frequently display pronounced differences in morphology hair and skin color bone structure and so forth that are almost certainly attributable to genetic factors and so testify to the relative integrity of the gene pool.

However owing to the relative ineffectiveness of phenotypic versus genotypic isolating mechanisms particularly when the phenotypic mechanisms do not have an immediate genetic focus continued geographic isolation is of greater importance in culturation than in speciation. Sympatric species are common enough (and there is some possibility that sympatric speciation occurs.[6] But genuinely sympatric cultures are rare if they exist at all. (A ghetto culture would not be a case in point for a ghetto is simply an instance of microgeographic isolation.) In consequence of all this cultures prove to be remarkably unstable structures: jet planes are as lethal for cultures as pesticides for species. (It is probably owing to the mobility of man and his ability to exploit diverse ecological niches that culturation though it serves to promote reproductive isolation has not led to speciation. A further factor may be that *homo sapiens* is simply slow in developing genetically based isolating mechanisms.)

Although art may be as it is said a manifestation of culture the aspect of aesthetic practice that we are concerned with the sophisticated form of behavior exemplified in the aesthetic aspection of a work of art such as Seurat's *Grande Jatte* is neither unique to nor universal in any culture. Even though the form of behavior in question has been very narrowly defined it is exemplified by people from very different cultures: by Japanese Swahilis Indians Americans Swedes Russians and so forth thus by people evidently belonging to different gene pools and having markedly different phenotypic repertoires. In consequence the fact that the behavior is exemplified by some and not others is not amenable to and cannot be given a purely cultural explanation.

This is not and should not be surprising. Although the production of works may be specific to a given culture as in Benin bronzes Navajo rugs Greek pottery the appreciation of such works is not. The appreciation of a work and the understanding of a discourse of a language stand in sharp contrast here. Those who are fond of drawing facile analogies between art and language tend to overlook the fact that understanding a discourse of a language is subject to drastic cultural restrictions whereas appreciating a work of art is not: one who cannot speak a word of Persian need have no difficulty in appreciating a fine Persian carpet. Language drives the wedge between cultures: it is the phenotypic isolating mechanism *par excellence*. To understand a discourse of a language one must require a detailed specific and extensive phenotypic repertoire that is a product of culturation: one must to that extent be acculturated. But no such extensive process of acculturation is required for the appreciation of a (nonlinguistic) work of art.

Neither is it surprising that only various people in various cultures perform such a behavior as the sophisticated aesthetic aspection of a work of art. Such phenotypic variability is the norm to be expected with higher organisms in highly varied environments. In considering the behavior of organisms it is generally useful to distinguish between communicative and noncommunicative behavior. The aesthetic aspection of a work of art is an instance of noncommunicative behavior: behavior in connection with a component of the environment.

For many species it is of considerable selective advantage to retain extensive flexibility toward components of the environment. Food sources come and go and so do competitors. Habitats change and an individual will encounter different substrates. Phenotypic flexibility rather than genetic precision is at a selective premium under these circumstances.[7]

However when a population of phenotypically flexible organisms manifests this flexibility in behavioral variations it may be difficult or even (at least at present) impossible to account for the occurrence of the variations. Since phenotypic traits are the expression of the interaction between genotypical and environmental factors to account for a given behavior would be to determine precisely which aspects of the behavior are attributable to genetic variations and which to environmental variations and of course to specify in so far as possible precisely what genetic and what environmental factors are in question. The difficulties in providing such an account should not be underestimated. For example "The black rat (*R. rattus*), a carrier of human plague, is a house rat in Europe and North America but a sewer rat in India, while the brown rat (*R. norvegicus*) shows exactly the opposite pattern".[8] No explanation for this difference is available. Again "the European pied fly-catchers that inhabit Spain differ from their more northern relatives in various details of 17 behavior patterns, though their habitats appear identical throughout Europe. No "reasons" for these differences have been made apparent".[9]

Why some and not others perform sophisticated acts of aesthetic aspection can for the time being be only a matter of speculation. Even so certain paths seem more plausible than others. "A New Zealand parrot, the kea, changed from an insect-eater to a predator of sheep when sheep were introduced".[10] The availability of provocative objects in the environment would seem to be a pertinent consideration. Surely humans are not less innovative and opportunistic than parrots!

I am inclined to suppose that anything that can be viewed is a fit object for sophisticated aesthetic aspection.[11] But this is not to deny that certain entities may provoke such attention more than others. To appreciate this fact one must realize that if each is intrinsically valuable the act of aspecting one work of art is not likely to be identical with that of aspecting another work in a different style or in a different tradition. Thus in aesthetically viewing a Botticelli *Venus* one attends to the moving contours but such an act performed in connection with a Titian *Venus* would not be intrinsically valuable would instead be frustrating owing to the absence of contours. Conversely although attention to pigmentation in the course of viewing a Titian *Venus* is of value it is not of value in viewing a work by Botticelli. Acts of aesthetic aspection constitute a large repertoire of behaviors each of which is a logically nonterminating act potentially of intrinsic value. But different entities make different demands on the viewer. Novel entities may call for novel behaviors for an extension of one's behavioral repertoire. And novel entities may prove provocative.

Entities are provocative if encounters with them give rise to an extension of an organism's repertoire of behaviors. Thus sheep proved to be provocative entities for the New Zealand parrot the kea. Milk bottles with foil caps have proven to be provocative entities for the English titmouse: encountering such entities in urban areas the titmouse has developed the behavior of piercing the foil with its bill and drinking the cream. It accordingly seems plausible to suppose that works of art such as Seurat's *Grande Jatte* have been provocative for many people in that encounters with such works have given rise to an extension of their repertoire of aesthetic behaviors: under the impetus of encounters with such works many perform what are for them novel acts of aspection. Instead of responding merely to the obvious physiognomic characters of the presented shapes they attend and respond to the various configurational aspects of the work as well.

The availability of provocative entities is I am inclined to think an important factor in the etiology of noncommunicative behaviors. But whether an entity is provocative depends on the behavioral repertoire already available and on the phenotypic flexibility of the organism. In America at present it is obvious that if one were to order the arts of music painting and poetry in terms of popular appeal and concern the order would be just that: music has a very wide appeal painting less so and poetry least of all. This is precisely what is to be expected if one attends to the factors of availability of provocative entities and available behavioral repertoires.

Music is virtually everywhere available supplied by portable radios TV sets records concerts guitar players and so forth. Furthermore the requisite behavioral repertoire for attention to popular music to be of some value is probably in part at least innate. Popular American music is dominated by the primitive beat of the tom-tom by rhythmic features calling for what is undoubtedly the most primitive form of aesthetic attention. But more important is the fact that in this culture there is no competing repertoire of behaviors in which musical sounds have any significant role to play. Musical sounds are distinctive in having an harmonic structure. Apart from minor occurrences in connection with door bells clocks push-button phones announcement of horse races and the like musical sounds occur primarily in pieces of music. In consequence no repertoire of behaviors other than those exemplified in attending to music is likely to be developed. But the development of an incompatible repertoire is precisely the factor responsible for the nonpopularity of poetry.

That poetry has less of an audience than painting is owing not to the availability of provocative entities but to the character of the available behavioral repertoires. Poems are probably as easy to encounter as paintings but the appreciation of poetry calls for a significant and conflicting extension of customary behavioral repertoires. Journalistic prose provides a paradigmatic representation of ordinary communicative behavior: in such prose the sound is totally subordinate to the sense. Indeed attention to sound as well as sense is virtually taboo in this society: witness the grimaces that usually greet puns. But a blend of sound and sense is the essence of poetry. Again ambiguity is a source of confusion in ordinary communication and is customarily avoided. But as Empson has superbly demonstrated in his *Seven Types of Ambiguity* it is an important and rewarding poetic device.

The neurophysiological base of vision is the visual cortex: this comprises one fifth of the human brain and is staggering in its complexity. Accordingly it is not surprising that the behavioral repertoires associated with visual matters are so vast and uncharted that no simple characterization can be given. But it is reasonably clear that unlike the case of music there are competing nonaesthetic repertoires of behaviors in which visual entities have an important role to play. We attend to looks appearances sights and so forth in a thousand different ways that have nothing to do with aesthetic matters. But unlike the case of poetry there is no single dominant repertoire that positively conflicts with an aesthetic attention to paintings. A society displaying the arts of music painting and poetry and in which the popular appeal of these arts was not in just that order is I think an extreme improbability.

The process by which an encounter with a provocative entity may give rise to an extension of a behavioral repertoire is difficult to detail. In considering the evolution of larvae of a New Zealand gnat from a fungus eating to a carnivorous type E. Mayr has conjectured that possibly the new behavior "started with a species which varied its fungus diet by scavenging, that is, by eating dead insects that had become stuck to the moist cave wall".[12] The behavioral repertoire of the gnats was then further extended step by step to encompass the catching of midges. What is in question is a gradual conservative extension of a behavioral repertoire. One can conjecture that the kea's change from an insect eater to a predator of sheep also involved a gradual conservative extension of its behavioral repertoire. Perhaps the sheep were parasitized by insects and at first the kea took to preying on the parasites.

To view the process by which an encounter with a provocative entity may give rise to an extension of a behavioral repertoire as involving a gradual conservative extension of the repertoire is to reduce but not at once to resolve the problem of how such an extension can take place. The dead insects presumably scavenged by the fungus eating gnats, were also provocative entities: what accounts for the first step being taken in a gradual conservative extension? Here one can conjecture that possibly failures of discrimination occurred which however proved to be adaptive. This is not an implausible view when minimal steps and lower organisms are in question. With higher and phenotypically more flexible organisms it is more plausible to postulate a genetically determined propensity at least for a time to experiment:

> Young birds generally try out a much wider variety of foods than those to which they eventually confine themselves. Such a trial and error period may be a typical stage in the growing up of the young in many species of animals (de Ruiter 1967). Whenever a novel habit was acquired by groups of Japanese macaques, it was invariably a young individual that took the lead (Kawai 1965).[13]

To compare a scavenging fungus eating New Zealand gnat encountering and for the first time ingesting a dead insect stuck to a moist cave wall with an art lover encountering and for the first time in 1948 viewing a work by Jean Dubuffet hung on the wall of the Museum of Modern Art in New York will undoubtedly strike many as altogether ludicrous. The fact of the matter is that each of these encounters exemplifies the same step in what is fundamentally the same biological process.

The history of art is the history of a gradual conservative extension of a behavioral repertoire under the impetus of encounters with provocative entities. Creative artists produce provocative entities: under the impetus of encounters with these works new forms of aesthetic behavior appear. A nice example is to be found in the emergence of nonobjective art earlier in this century in the works of Mondriaan Kandinsky and others.

Although nonobjective decorative art has been around for centuries nonobjective easel paintings are a 20th century innovation. If one turns directly from a Botticelli to a Mondriaan there appears to be an impassable gulf between them. Indeed if *per impossibile* Botticelli were to have viewed a Mondriaan he would not have known what to make of it. For the viewing of a typical Mondriaan to be of aesthetic value a particular act of aspection is called for: one must attend to the stability of the picture plane to the spatial interaction of the black lines with one another with the edges of the canvas and with the picture plane. No such act can be performed in viewing a Botticelli. There were nonetheless stepping stones available in the Western tradition with which this gulf was bridged.

Viewing a Poussin calls for an act of aspection different from that to be performed in viewing a Botticelli: three dimensional spatial organization comes to the fore assumes far greater importance demands careful attention. Yet the step from Botticelli to Poussin retrospectively at least does not appear staggering. (And this gap too was nicely bridged by the intervening tradition.) And from Poussin to Ingres and Ingres' drawings in particular the change was slight (which is not surprising since Ingres was a great admirer of Poussin). From Ingres' drawings to Cézanne's watercolors and Cézanne's apples was again not a prodigious leap. To view an Ingres drawing one must attend to the spatial interaction of the contours to the careful delineation of bumps and hollows. The same is true in viewing a Cezanne watercolor but in Cezanne the image had already begun to dissolve into spatial relations with a correlative alteration in the act of aspection to be performed. From Cezanne to the Cubists to the analytic cubistic works of Braque and Picasso and to Mondriaan was then but a hop and a skip and a jump.

There is an important difference between the fungus eating gnat ingesting a dead insect and an art lover viewing a Dubuffet. Becoming carnivorous increased a gnat's relative fitness (since the carnivores evolved from the fungus eaters). But there is (at the moment) no reason to believe that becoming a devotee of Dubuffet will increase anyone's relative fitness. So one might be inclined to argue that the encounter with the dead insect and the encounter with the Dubuffet do not exemplify the same step in the same biological process. This would be a mistake.

An increase in relative fitness is not itself a biological process but a sometimes product of or a step in a biological process. Consider a population of fungus eating gnats in an environment in which fungi are plentiful and insects scarce. Becoming carnivorous would then decrease a gnat's relative fitness. Or if both fungi and insects are in sufficient abundance there might be no effect on relative fitness. But in any case the biological process of a gradual conservative extension of a phenotypic repertoire under the impetus of provocative entities remains the same. (One should note in passing that the related process of operant conditioning also does not directly determine relative fitness: it too may lead to adaptive or maladaptive or non-adaptive behaviors.)

The relevant difference between the gnat's and the art lover's behavior is that the gnat was in quest of food the art lover of a work of art. Eating and viewing each have a pay-off but not the same pay-offs. The primary value of eating is obviously utilitarian: it is a means by which organisms are supplied with energy. The primary value of an act of aesthetic aspection is nonutilitarian: if it is of value it is of value regardless of whether it is also a means to some further end. But the fact that the value of an act of aesthetic aspection is intrinsic does not preclude the possibility of its having some biological value as well.

Does it have any biological value? Since the behavior is biologically speaking today's innovation of *homo sapiens* there can be no historical evidence that it is adaptive. And unfortunately virtually nothing is known at present of the underlying neurophysiological mechanisms. Eating behavior is directly under the control of certain centers in the hypothalamus and this control has been demonstrated. One can conjecture vaguely that aesthetic behavior like emotional matters must have something to do with the limbic systems but there are no specifics available. Conceivably aesthetic behavior could be attributable to a nonadaptive pleiotropy testifying to nothing more than the phenotypic flexibility of *homo sapiens*. But conceivably not.

There is a sense in which a world of utilitarian values without any nonutilitarian values would make no sense at all. For if *a* is of value only because it leads to *b* then *b* must be of value but if *b* has only extrinsic value then *b* is of value only because it leads to *c*. And so on. If there is no *i* such that *i* is intrinsically valuable then not only is the value of *a* ultimately illusory but one is caught in an endless futile pointless process. Traditional biological theory has been and contemporary theory still is focussed only on the struggle for existence. In this struggle all values seem to be extrinsic: one struggles to eat to gain energy to reproduce have offspring who will struggle to eat to gain energy to reproduce have offspring and so on and on and on. When one is engaged in this struggle there may be no time for the enjoyment of aesthetic values. A snowflake on a branch may invite aesthetic attention but one can hardly accept the invitation while being savaged by a pack of wolves. This may be why anthropoid apes seem to display no aesthetic behaviors.

When Thomas Aquinas was faced with the problem of what the beings in heaven were to do for all eternity he had the happy thought that they would contemplate the Deity thus perform a logically nonterminating intrinsically valuable act. We are no where near achieving a heaven on earth but the technological achievements of *homo sapiens* have if not ended the struggle for existence at least established for some some significant pauses in the fray. It is in these interstices of the struggle that the locus of intrinsic value is to be found.

The human genotype surely does not determine but just as surely allows for the occurrence of forms of behavior to occupy these pauses in the struggle for existence behaviors which make the struggle worthwhile. It is essential to realize that genetically based capacities each of which has an explicit utilitarian value may combine to yield a nonutilitarian value. First self-awareness has a basic role to play in the development of a recursive technology. One is not apt to use a tool to make a tool unless one is aware in using a tool that one is using a tool. Self-awareness is similarly implicated in the recursive structuring of linguistic systems.[14] Secondly the capacity to respond to physiognomic characters may be of utility in dealing with possible threats: facial expression movements shapes are seen as threatening friendly and the like. This is no doubt a primitive device of dubious utility but here one must remember that the human genotype is a primitive evolutionary product possibly not much changed since Cro-magnon man (*Homo sapiens sapiens*) invaded Europe some 35 000 years ago. And finally a capacity to appreciate the structure of a complex entity is of evident utility in predation. Such a capacity would seem to be relatively widespread throughout the higher organisms. If in some pause during the struggle for existence each of these capacities is exercised simultaneously under the impetus of an encounter with a provocative entity the result may be the performance of an act having a nonutilitarian value an intrinsic value an aesthetic value. Aesthetic value then is a pot of honey found along a thorny biological path while the wolves are at bay. It is not the only such pot: there are many paths to euphoria. But it is an important one.

What then is the biological value of intrinsically valuable aesthetic behavior? Simply to be. And can sociobiology contribute to an understanding of contemporary aesthetic practices? Of course it can and anyway if it cannot then nothing can.

VII

HOMO SAPIENS TELEVISUS: A SOCIOBIOLOGICAL MEDITATION ON TV

In this culture I will not speak for any other we are watchers of television. We sit alone or huddled together in darkened rooms staring at the tube. (Big Screens are on their way but today is still the day of the tube.) Despising it we call it the boob tube the groove tube the idiot box but we still sit and stare. The TV set is a cultural product a technological achievement of the twentieth century. Yet the images it purveys have primeval roots: Cro-Magnon Man *homo sapiens sapiens* would have been a TV fan.

Cro invaded Europe some 35000 years ago. I don't know where he came from. He didn't have a TV set to tune in but he was tuned in to himself. He saw himself reflected in sometime quiet pools stared at the reflected image recognized himself. No mean accomplishment but ordinary for such a primate. Rhesus monkeys when they see the aggressive display of a Rhesus displayed on a TV screen respond aggressively and never ask 'Is what I see me?' But we do and Cro did.

How did Cro tune in to himself? Some mutation was plugged in when his grand daddy's head grew big. An inelegant hypertrophy in the neighborhood of the pituitary gland. It doesn't matter how. He did. Then now and then Cro did things like gathering and hunting. Pursuing game was no easy sport. Failure and success mattered much and were attended by feelings of grief or rage or delight. Tuned into himself he executed plans and attended to himself executing plans. This is important.

Cro was not like a pair of cheetahs whose plan of attack is executed with splendid unselfconscious aplomb. A herd of Thompson's gazelle three hundred yards off across the grassy savannah is spotted. The cheetahs walk. In plain view calmly steadily forward. The nervous herd grows more nervous still. The cheetahs still walk. The tension builds. (We'll not turn our attention here to catastrophe theory though here one could.) Then one of three is likely to be: the herd as one wildly mobs the cheetahs who then turn and run. Or two: the herd as one will run away while the wily cheetahs follow swiftly while still waiting. For three: one breaks away from the herd and the pair of cheetahs separate to encircle their prey cut off retreat. But though cheetahs beautifully execute this plan no cheetah is aware in executing this plan that it is executing this plan. Because unlike Cro no cheetah is ever tuned in to itself. Cro did what he did and was aware in doing it that he did it.

Cro would hunt all day but at night in his cave having no TV to tune in he tuned in to himself. Did Cro dream? Cro did dream. (You don't have to be a Freudian to believe that Cro had dreams: Freud didn't invent dreams.) Cro would dream and see himself doing what he sometimes did or might do or never did. He had an active visual cortex a lively reticular activating system: Cro was a big head and altogether capable. Did Cro sometimes when awake rehearse in his head what he did before he did it? Not having the benefit of TV film playmakers someone to stage scenes for him he was thrown back on himself stuck with his own resources. (Who's better off? Cro or today's kid?) Cro was forced to develop his capacity to envisage a course of action the execution of a plan forced to develop his imagination. Cro imagined scenes.

How come he imagined scenes? I don't know. But he did. Maybe AlterCro didn't. Then AlterCro didn't make it the way Cro made it. Because Cro maybe caught more bison because Cro's visions must directly or indirectly have been selected for. (Perhaps indirectly because possibly some pleiotropy was involved and the visions were merely a concomitant factor.) But this seems plausible: rehearsing scenes in your head can help because when you're chasing a bison there's no time to think and you don't want to think lest the native hue of resolution be sicklied o'er with the pale cast of thought. It doesn't matter why Cro took to imagining scenes but he did and if AlterCro didn't then AlterCro didn't make it: his genes were wasted. None of his kids are around today. Apart from idiots and the like any *homo* can imagine scenes.

Alone in his cave at night Cro imagined scenes. Cro took to composing scenarios. And the first theme was success. Because if he could succeed he would eat and grow strong and get a woman and reproduce. Cro didn't know it but success meant an increase in fitness and Cro was caught in a violent struggle for existence. Cro imagined scenes of success. At first just simple scenes. He imagined himself eating. Good! He imagined himself eating. Good. He imagined himself eating. Ok. He imagined himself eating. No. This time it was a big bison he was eating not just eating. Good. He imagined himself eating a big bison. No. This time it was catching and eating that big bison. For even Cro could be plagued by the *tedium vitae* of an imagined world. And to catch and eat the big bison that almost but not quite but almost got away was still better and added spice to the imagined dish.

Not only did Cro rehearse plans in his head but in his head he envisaged the likely upshot of this or that course of action and in doing that he experienced anticipatory grief or rage or delight. Anticipatory grief is only the shadow of a grief cast forward but it is grief of a sort. Cro was sometimes afflicted with it. He didn't like it. Sometimes he imagined himself tracking a bison being himself trapped and gored. A bad scene for a bad dude! Cro didn't like the scene of Cro being gored to death. So Cro worked out the archetypal TV scenario for himself: that of the invincible tested and proved. And whereas Cro tuned into himself exercising his genetically based capacities for self-awareness planning imagining the execution of a plan and experiencing grief or rage or delight we tune in to TV to view Cro's first scenario detailed scripted twisted turned a thousand thousand different ways but always the same.

In Cro's first scenario Cro is invincible and so allied to Superman Mr. West and Spider Bionic Man and Jaime. Of course no one is invincible. Everyone knows this and knowing it knows that the Phantom Rider mounted on Silver is invincible. So this titilating dialectic is played over and over and over and over again for no one is invincible and He is invincible and we view His exploits with trepidation but not trepidation for we know He is invincible which if the action and when the action is fearful we remind ourselves never having forgotten it but letting it sit safely in the corner of our awareness in the cellars of attention.

For I tell you that we are as Cro would have been a fan of spaghetti westerns and we stare as Cro would have stared at the tube hour after hour watching this theme unwind its derring do suspense somehow not undone by foreknowledge but rendered ever the more delightful in innumerable forms but always recognizably what it is in the tribulations of Jeannie's master or Samantha's man or when Malloy is caught in the crossfire of a stakeout or Rockford is about to be wasted when Quincy or Barnaby or even Hogan is at a loss and Jed is about to be swindled and Lucy to lose we know it is important that we know and we do don't we that Charlie's Angels will triumph as surely as Secret Agent man the Saint double o seven Cain's Kung Fu and the Hulk's indomitable rage. With a billion biological eyes we comfortably uncomfortably attend to the trials of the invincible secure always knowing and in the knowledge that the verdict will be: the Road Runner will not be caught the Impossible Mission will succeed Pauline will survive her Perils while baggy panted duck gaited derby topped cane swirling figures will ride off into forever sunsets as Silver snorts and prances.

The invincible tested and proved was Cro's first scenario. But unlike most of us Cro was soon bored with so simple a scene. Cro hit upon his second scenario which in truth was no more than a variation on the first. It was a slight refinement: he hit upon the idea of perfection. He imagined not simply the invincible tested and proved but the invincible perfect tested and proved. So he imagined the perfect hunter the perfect bison the perfect spear the perfect burin. But Cro was limited. His was a primitive impoverished culture: there were fewer roles to play.

Today's TV script writers in all their wisdom drawing on the current cultural scene display a passion for perfection of a kind. James Bond is our perfect secret agent. Dr. Welby the perfect family doctor. Gannon the perfect surgeon. Malloy the perfect squad car cop. Sergeant Saunders the perfect combat sergeant. Joe Friday the perfect police sergeant. Bionic Steve is the perfect astronaut. The Father who knows best is the perfect father. Perry Mason the perfect lawyer. Ironside the perfect police chief. Charles the perfect prairie farmer. Lou Grant the perfect editor. Inspector Erskine the perfect FBI agent. Matt Dillon the perfect Western sheriff. Jim West the perfect Western agent. Gene and Roy the perfect cowboys Dale the perfect cowgirl. Chrissie the perfect dumb blonde. Mary Richards the perfect girl in the office. Andy Taylor the perfect small town sheriff. Aunt Bea the perfect Aunt. Barnaby the perfect elderly investigator (harking back to radio's Mr. Keene the kindly old perfect tracer of lost persons (hurray for Bob and Ray! a lost bright light in our cultural blight)). Is Samantha the beautiful witch the perfect wife? Anyway Lassie is the perfect dog Silver the perfect horse.

Being a primitive being Cro like all of us was susceptible to respondent and operant conditioning. But as his own script writer he was not prone to self-exploitation. Unlike us being free to choose he would not have chosen to interrupt his visions with a sudden importunate message to select the right tampon compare pure Sweat with the leading roll-on squeeze Charmin against his indelicate anus. Cro was his own supplier but we are pandered to to be manipulated by Exxon & Co. And our genetically based susceptibility to respondent and operant conditioning is investigated daily on Madison Ave.

Confusion of a symbol with the thing symbolized is endemic in our culture. Each morning all over America little kiddies celebrate and perpetuate the conundrum of pledging "allegiance to the flag and to the republic for which it stands": allegiance to a republic is perhaps possible: "allegiance" to a piece of cloth must puzzle even the most patriotic will. TV commercials are dedicated to the proposition that symbols and the symbolized form a sacred union. See Catherine de Neuve a bottle of Chanel No. 5 beautifully sprawled across the hood of a limousine: Buy perfume this woman! Are you Ugly? Paint your face as Cheryl Tiegs paints hers and you will won't you look like Cheryl baby or maybe Miss Piggy. Don Meredith drinks Lipton Tea. Don't you want to be a Dallas quarterback? Hop to it: Buy some Lipton Tea! Cougar Thunderbird Mustang: lovely creatures lovely names: buy the products so-named! (What's in a name? Much: Schmidt's beer is doomed never to rival Schlitz in sales: for only the latter has a light Viennese lilt. Marlboro out of Marlborough (sired by Madison Av) has spent millions to overcome its initial ladylike stance which is why we were shown innumerable hard-nosed western boys puffing away in the wilds. And what inept namer named the Edsel?) Look at this fully equipped racing striped bucket seated gorgeous long haired silky racing mirrored model: Buy this (plastic) beauty! (What amateur ran the add showing Volvo owners surviving crashes which sent sales crashing? Thinkin bout cahlamaties doan wanna make no one buy no car.)

Certainly Cro took to symbols. But was he taken by them? Possibly he was. Probably he was. Possibly he became a symbol chauvinist pig: conditioned by his own images he came to prize symbols more than realities: to him the real became unreal or might as well have been unreal or was as good as unreal if it had not loomed larger than life in his inner eye if it was not symbolic. Poor Cro! The big grey pig he caught and ate that filled his belly became less important to him than the little white one that got away. The small crude flint he stole from AlterCro mattered more than the one he shaped on his own. Being free to imagine scenes he imagined that he was free: instead he became the captive of his own imagination. He became a human being. He went insane.

Cro of course was familiar with violence. But his symbolic schemes were limited. He perhaps scrawled symbols of aggression on cavern walls but he lacked the contemporary cultural means of expression. Cro imagined scenes with spears and clubs: in today's TV a giant cultural step forward has been taken: the club has been replaced by the gun (aided and abetted by rubber truncheons black jacks brass knuckles grenades tear gas scoped rifles shot guns bombs and not to forget underwater spears accompanied by handcuffs straightjackets deadly lasers phasers and what have you).

(Monday April 21st 1980 in Gatorville Florida: how's de gun doin tonight? 7 : 00 pm Jim Rockford: he doan carry one but one is used on him all de time. Home Box Office: "Johnny Got His Gun": the tragic story of a WWI basket case. 7 : 30 Mash and the Korean War. 8 : 00 Movie: "Winchester '73'" (hey de gun is definitely dere!) 10 : 00 Big Battles: "The Battle of the Pacific – The Setting Sun." 11 : 00 Western: "Bandits." 11 : 30 "Little Caesar": a small town hood rises to the top of the underworld. 11 : 50 even Barney Miller totes one. 12 : 00 am "Streets of San Francisco" BANG! BANG! Movie: Western: "Warlock": a gunfighter hired by the town etc. 12 : 20 am Police Woman. 1 : 00 am Movie: "Trip with the Teacher": ... a nightmare of terror. 1 : 15 am "Murders in ..." Basta! TV is Gun Ho!

The best estimate of the number of guns in private hands in this country is ninety million; but guesses supported by some data range from fifty million to two hundred million. Each year an additional three million guns are purchased, and very few older guns become unusable. Nor does this massive weaponry lie idle. In 1968, by means of guns there were committed 8,870 murders, 64,950 aggravated assaults, and 99,000 armed robberies. Guns were also used in that year in over 10,000 suicides and in over 2,500 "accidental" deaths. The estimated total of nonfatal injuries was 100,000. Guns were used as the murder weapon in 96 percent of the 475 killings of policemen during the period 1960–68. Moreover, since the beginning of this century some three-quarters of a million people have been killed in the United States by privately owned guns, 30 percent more than in all the wars in which this country has been involved in its entire history. In sum and in short, the populace is armed with a dangerous weapon.

> Yet it is a curious paradox that while the possession of a relatively innocuous drug whose potential for harm, if it exists, is limited to those who take it voluntarily is a criminal offense subject to draconian penalties, the possession of guns is subject to scarcely any control in most states.[1]

It's ok to buy a gun and booze but not grass. Sure guns and drunks are a menace: a marijuana smoker is likely to be a quiet harmless bore. Yet there's no paradox here: the symbolic is what matters to the symbol minded: a man with a gun in his hand is a man. That's what meets the symbol fancier's eye. The joint smoker coke sniffer is limp with a glazed look slack stance. He's turned on but his image turns off. While a man with a drink in his hand (depending on the drink) is suave debonair polished virile hearty. Not reasons but images base and debase our public policy.

TV epitomizes the symbol minded mania that dominates the human race. TV is today the supreme purveyor of images and the image is all that matters. But the images are old. Old. In Shakespeare's *Hamlet* Hamlet dies and Rosencrantz dies and some weep for Hamlet but who weeps for Rosencrantz? Prince Hamlet is a prince and Rosencrantz is a minor character. Rosencrantz isn't even a real minor character: he's an imaginary minor character a fictitious person. But then Prince Hamlet isn't real either: even so he's an imaginary prince. Ham and Rose have their counterparts their twins everywhere in fiction: in plays novels ballads and in sagas certainly in epics. Ham has dominated our theatre for a thousand years upstaging everyone yet Roses are everywhere. What would a Western Horse Opera be without innumerable Roses falling from saloon roofs countless red Roses biting the dust Roses in a circle being smitten by arrows chased by sheriffs hit by lead being always slower on the draw than Ham?

Roses are always dying and who cares? It won't be reported on the late evening news or if it is it is without pictures. The First Ambassador entering the scene finding Horatio with the dead says:

> The sight is dismal;
> And our affairs from England come too late:
> The ears are senseless that should give us hearing,
> To tell him his commandment is fulfill'd;
> That Rosencrantz and Guildenstern are dead:
> Where should we have our thanks?

Horatio has other things on his mind: he is not over concerned with the demise of Rose and Guildy. Neither is the audience. Hamlet has just died: "Good night, sweet prince, And flights of angels sing thee to thy rest!" Horatio's words are still echoing in our ears: to weep for Rose at this moment would surely and merely mess matters up. Drama calls for the dramatic. If someone plays prince then won't someone have to play courtier? That's the way it's always been.

Why weep for Rosy? The play isn't much concerned with him. If he were cut from the script it needn't matter much. Things could go on much as they had been going on. Whose attention is focussed on him? Whose imagination does he capture? He sets no one thinking makes no one wonder. Does he matter? In fiction that's the way it often is: major figures matter minor ones don't.

Is it like that in real life? Am I a minor character and Ronnie Reagan a major? I am real but then maybe so is he. Is he a Ham and am I a Rose? There's something to be said for that. Certainly he's a big wheel in the political machine in which I am at best a bit of grit. From that point of view his existence matters and mine doesn't. But there are other points of view: he's not the focus of all eyes. When I come home my cats look to me not him. When I visit a zoo and stand staring at the aoudads they stare back at me.

When it comes to the real unreal Ham and Rose there is only one point of view: that's the one supplied by Shakespeare: Rose being unreal a fictitious person lacks reality substance when viewed from any perspective other than that supplied by his author. Offstage Rose keeps no cats is cared for by no one does not exist. So the death of Rose doesn't matter to those who view him as what he is: a minor character. (Not to digress: but do we view Rose as a minor character or as a minor character in a play? As a minor character surely not as a minor character in a play. Nor do we view the play as a play unless we're playing critic.)

The human race is being done in by a fancy for the symbolic graced by an amazing nurtured perhaps by a carefully cultivated stupidity cupidity ignorance. To fancy the symbolic to take an aesthetic stance in connection with works of art of drama of literature is right and sensible and sensitive. It is fine proper and good. In short: it is just dandy. But to view the affairs of real life that way to fix on attend to be moved by symbolic aspects when confronted with war pestilence and famine is to invite the cataclysm of the apocalypse: it is insane.

Are we insane? By and large we are. Our society certainly is. (Isn't that obvious? Just look around you! No! no! First open your eyes!) Or if not positively insane then we are dupes naifs poor innocents victimized not by any grand world conspiracy but by our own over heated imaginations. For we allow symbolic aspects to determine our behaviour attitudes in areas of life in which such factors have no rational role to play.

It is not comforting to think that no truly ugly one could be president of these United States one who like a leper had lost his nose who presented himself barefaced noseless. What's a president need a nose for? Was it his fine wide grin and Daddy Warbucks like look that led to Ike's being elected? Consider Cally: charged with killing unarmed children men and women. A heinous offence yes but less so than annihilating an urban center. The pilots who dismissed Hiroshima to atomic hell have not been crucified as Callys. They wore other masks: Cally was cast as a heavy in a Hollywood western: the bomber pilots as anonymous cogs in a vast war machine.

50 000 Americans or more were killed in recent years in Viet Nam. Students demonstrated police rioted students rioted buildings were

seized students clubbed by construction persons shot by militia: all this occasioned by and in connection with protests about the war. During the same period some 500 000 Americans were slaughtered on our nation's splendid highways. Dere aint bin no riots bout dat. (There is no technological problem in building a collision proof car: Cornell Univ. designed one years ago: crash head on into a stone wall at 60 mph and walk away unscathed! The image wasn't sexy: the car lacked a sleek powerful lethal look.)

And *Ave*! Mission Apollo first class prime time TV spectacular at a budget of some 25 billion dollars high even for a Cecil B. production. "One giant leap for mankind" is what the man on the moon said: but space travel agencies aint making big bucks: seems like there's no place in space space worth going to and no time to get there.

Our wise leaders willing to spend 25 billion on the mighty dramatic derring do of a Moonwalk live on TV AND IN COLOUR come all over solemn prudent and budget minded when it comes to such undramatic unsymbolic trivia as the plight of 650 000 American Indians.

> Fifty thousand Indian families live in unsanitary, dilapidated dwellings: many in huts, shanties, even abandoned automobiles. The unemployment rate among Indians is nearly 40 percent – more than 10 times the national average. Forty-two percent of Indian schoolchildren – almost double the national average – drop out before completing high school. Indian literacy rates are among the lowest in the nation; the rates of sickness and poverty are among the highest. The average age of death of an American Indian today is 64 years; for all other Americans, it is 70.5.[2]

The misery of the American Indian has no symbolic appeal: year after year we've seen red skins in our flicks and on TV being shot down by brave pioneers. Anyway they're picturesque in their poverty. Poor people are boring: subjects for low comedy not high tragedy. Beaming at the first moon-walk a moon struck president spoke of the "national pride" in that achievement: no one be doin any talkin bout any "national pride" in our many green ghetto where the great rats play. But as a good Republican said if you've seen one ghetto you've seen them all.

Two American soldiers shot along a Korean border: at once a major task force is mounted and dispatched to the area. Every day every single night Rose is rolled in Toledo mugged in Minneapolis raped in Chicago battered in Cleveland stabbed in Los Angeles murdered in Detroit. "Mr. Prez: how come you doan moun no task force to Detroit? We be hurtin bad. An Mr. Brinky and Mr. Chanceliar why doan yo say on de TV news dat Rose was murdered in Detroit?" Well you know: TV's concerned with somebody who is somebody. He was nobody.

How can somebody be nobody? He can. In this world he can. We live in a dramatic lunatic asylum where the keepers are even madder than the inmates and where keepers and inmates all sit staring at the tube waiting watching waiting to see mushrooms grow in the sky. See it dead on TV.

VIII

ANYTHING VIEWED

Look at the dried dung!

What for?

If I had said 'Look at the sunset!' would you have asked 'What for?'?

People view sunsets aesthetically. Sunsets are customary objects of aesthetic attention. So are trees rocks wildflowers clouds women leaping gazelles prancing horses: all these are sometime objects of aesthetic attention. But not everything is: not soiled linen greasy dishes bleary eyes false teeth not excrement.

Why not? It's not because they're unbeautiful or even ugly. Beautiful things are no problem for a rambling aesthetic eye but not all objects of aesthetic attention are beautiful: Grunewald's *Crucifixion* isn't neither is Picasso's *Guernica*. Breughel's rustics aren't lovely. The stark morning light in a Hopper is powerful but it is not beautiful. Not being beautiful needn't matter.

These unbeautiful objects are works of art. By chance some objects of aesthetic attention have been naturally produced. For the rest: they are products of art.

What is a work of art? Something fit to be an object of aesthetic attention. Most likely nowadays (now that didactic art is largely dead) something tailor-made for the purpose designed to be just that. If you want to attend aesthetically to something fix on a work of art as your object: that's the way it's thought to be. Is a work of art the paradigm of an object fit for aesthetic attention? What does a work of art have or lack that dung doesn't?

What is a work of art? Not everything. Leonardo's portrait of Ginevra de' Benci is. A mound of dried dung isn't. Nor is an alligator at least a living gator basking in the sun on a mud bank in a swamp isn't. A reason they are not is plain: nothing is a work of art if it is not an artefact something made by man. A gator basking a mound of dried dung are products of nature made or produced by natural forces. Not being made or produced by men they are not classed artefacts. Not being artefacts they are not classed works of art. Such is a common or the common if there is anything that is the common conception of a work of art.

Most likely there is no such thing as the common conception of a work of art: these are vague ill-defined notions. And some say that some objects that are not artefacts are nonetheless works of art. That needn't concern us: undoubtful examples of works of art are all that are wanted here and now and these are easier to come by when one considers artefacts rather than nonartefacts.

When one looks at a gator basking a mound of dried dung is one at once cognizant of the fact that not one or the other is man-made? And does such cognizance at once preclude all possibility of aesthetic attention to the gator basking the mound of dried dung? Though the gator basking is not man-made it is (to invoke the shade of Paley) remarkable in design and structure. By no stretch of the imagination can it be imagined to be less detailed rich intricate in design less complex in structure than an artefact. Given the present state of technology there's no way anyone can actually make a gator basking. But making a mound of dried dung is easy. Conjure up this image: a field in which there are two virtually identical mounds of dried dung. One was and the other was not man-made. Would that fact render the latter less accessible than the former to aesthetic attention?

Imagine this: that the Henry Moore statue at Lincoln Center was in fact not an artefact by Moore but a naturally formed that is nonman-made object found in a desert and transported to Lincoln Center. Would that matter to an appreciation of the statue? Yes enormously. Knowing that one's view of the object would be restructured: one would not in looking at the work look at it as a work. One would not look for manifestations of craftsmanship. One would not look for and see signs of the sculptor's hands: there would be none. But the object would still have shape form mass and balance. The various parts of the object would still be in the spatial relations they are in. The solidity of the volumes would remain unaltered. Nor would the expressive aspects of the object be seriously impaired if impaired at all by its lacking the status of an artefact. It would still possess those physiognomic characteristics which serve to make it an imposing impressive work. That it was not an artefact would not indicate that it was not a fit object for aesthetic attention.

That something is not an artefact does not suggest let alone establish that it is therefore unfit to be an object of aesthetic attention. And unless one has a compelling narcissistic obsession with the marks of men's endeavours one can view things in the world aesthetically without being concerned with or inhibited by their lack of status as artefacts.

If a work of art is a paradigm of an object fit for aesthetic attention it is not owing to the status of a work of art as an artefact. Not that just any artefact is classed a work of art: a garden rake a screwdriver a green paper plate are not though they are undoubtful examples of artefacts. What if the paper plate were on a pedestal displayed as a piece of sculpture? Would it then be classed a work of art? By some. Not by others. Even so: if one wanted an undoubtful example of a work of art wouldn't one prefer Leonardo's *Ginevra* to the paper plate? An undoubtful example of a work of art is a hand-made work a product of an art a craft: it is an artefact the production of which called for considerable and unmistakable craftsmanship. Look at Leonardo's *Ginevra*: that the craftsmanship displayed is remarkable is obvious. (And that is not belied by the fact that one may wonder whether the portrayed slight strabismus is rightly to be attributed to Ginevra herself.)

This exquisite portrait is incomparably more beautiful than any reproduction can suggest. The marvelous sense of atmosphere surrounding Ginevra, the harmonious unity of landscape and figure, and the incredible delicacy with which minute details are rendered can only be appreciated in the original painting.[1]

Reproductions rarely capture the quality of a work of art of an exquisite and refined craft. That a work does not lend itself to easy reproduction however may be owing either to its being remarkably ordered (so to speak) a product of great craftsmanship or to its being a clear manifestation of entropy. Leonardo's *Ginevra* would be difficult to copy and so would one of Pollock's typically dribbled pieces: to smash an egg is easy but to replicate the appearance of the smashed egg in all perceivable details may be impossible.

A display of craftsmanship may on occasion facilitate aesthetic attention to an object. The lack of that display in no way indicates that an object is unfit for such attention. Consider a typical work by Piet Mondriaan: one of black lines and white ground. Such a work displays virtually nothing of the painter's craft rightly so-called: a tolerably steady hand an ability to apply masking tape judiciously is about all the the technical skill required to produce it. Or to reproduce it: a perfect copy would be a matter of a few hours work at most.

That works of art may be artefacts that they may be skillfully hand-made objects here doesn't signify. Figuratively and on occasion literally speaking works of art are framed objects. It is that more than anything else that makes them plausible paradigms of objects fit for aesthetic attention. But both the efficacy and the necessity of a frame are something of an illusion.

Works of art are framed mounted hung illuminated displayed exhibited. The object is supplied with a milieu an environment a background. Presumably all that facilitates aesthetic attention to the works by those concerned to appreciate them. The basic idea would seem to be this: a person p performs certain relevant actions a in connection with a work of art an entity e under conditions c. The entity e is supposed to be of a kind or character to facilitate and make valuable the performance of *a* by *p* under *c*. If so *e* is then a fit object for aesthetic attention. And what if *e* is dried dung? Then the performance of actions *a* by person *p* under conditions *c* in connection with *e* the dried dung is supposed to be neither facilitated nor rendered valuable by the dried dung. Hence the dried dung is not supposed to be a fit object for aesthetic attention. But obviously all this depends on the person *p* the actions *a* and the conditions *c*.

Aesthetic value is as it were a cooperative affair. If attending aesthetically to an object is worthwhile then the object contributes its presence and possibly the conditions under which one attends to the object contribute their share while the person contributes his: what is wanted is an harmonious relation between the person and the object. It is never the case that such harmony depends solely on the contribution of the object. For despite its presence the conditions of attention may be infelicitous: who could enjoy viewing Klee's *Twittering Machine* while being tortured? (Perhaps a *roshi*.) If both object and conditions make their contribution something about the person may occasion a difficulty: a color blind person may be cut off from an appreciation of a Matisse nude and so conceivably could one psychologically disturbed about sexual matters.

To say of something that it is worth attending to aesthetically is to speak in an abstract way. For in so saying one abstracts from reference to persons actions and the conditions under which the actions are to be performed. On occasion this abstract way of speaking is somewhat fatuous. A case in point: 'Michelangelo's Sistine Chapel murals are worth viewing.' Presumably these are great works of art. Theoretically the viewing of these works is aesthetically worthwhile. In fact it is not. It would be worthwhile if the works were not where they are if the conditions of viewing were altered for example if the Chapel were turned on its side. Where they are high up and almost out of sight they are for all save presbyopes virtually inaccessible to the performance of any relevant aesthetic action. Viewing them is literally a pain in the neck. One can recline on a bench or the floor (if the guards permit and the spectators don't trample) but that position is not conducive to aesthetic attention. Here one should keep in mind the illusion of the full moon on the horizon: the apparent size of the moon is radically reduced by turning one's back to it bending down and viewing it between one's legs with one's head upside down. Evidently the positions in which one views things can serve to alter the apparent size of the things viewed. (It is said that Frank Lloyd Wright hated paintings: that would account for the sloping floors and tilted perspectives of the Guggenheim Museum which serve effectively to sabotage any delicately balanced work.)

A work of art is supposed to retain its identity from frame to frame wall to wall room to room: those who suffer from inept framers know how silly this view is. Seurat took care at times to prepare and paint his own frames. But he could do nothing about the walls floors company his works were forced to keep. Conversely is there any doubt that dried dung displayed by the lighting engineers of the New York Museum of Modern Art could prove to be a fantastically intriguing aesthetic object? With appropriately placed lights and shadows walls of the right tint in the right position of the right height carefully proportioned pedestals anything at all that could be displayed could be a fit object for aesthetic attention.

Would it be the dried dung or the dried dung under special environing conditions that would be a fit object for aesthetic attention? Certainly at least the latter is obviously true and I think also the former but let's focus on the latter for the moment for that's the way it always is anyway with any work of art. Works of art such as paintings and pieces of sculpture are best thought of as scores awaiting realization in actual performance. Viewing a yellow version of Josef Albers' *Hommage to the square* displayed in a yellow frame on a yellow stuccoed wall would be like listening to a Rossini overture performed *con sordini* with all instruments muted.

To say that an object is fit for aesthetic attention is not simply to say that there are or could be environing conditions under which the object would be worth attending to aesthetically. That seems plainly true (to me anyway) and not surprising in the light of twentieth century art and techniques of display. In saying that an object is fit for aesthetic attention one is saying much more namely that the object can be attended to and is worth attending to aesthetically in that such attention to the object is worthwhile and if it is not that it is not is attributable either to interference by the conditions or to something about persons or their actions.

When attention to an object is not aesthetically worthwhile it may be uncertain what the lack is attributable to. If aesthetic attention to a floating clump of seaweed was not worthwhile that may be owing to the fact that while contemplating the clump one was being savaged by a school of sharks. Here conditions may fairly be said to have interfered. But if on a cold dank winter's day in Venice one finds the contemplation of a Tintoretto in a dim unheated church not aesthetically worthwhile is the lack to be attributed to the conditions under which the work is viewed or to a failure of concentration on the part of the person?

As the character of the objects attended to vary the character of the actions the conditions and the requisite qualities skills and capacities of the person may also have to vary if attention to the objects is to be aesthetically worthwhile. Demands made on a person are absolutely minimal in the appreciation of the popular art of his own culture: soap operas rock and roll comic strips western flicks. No special knowledge is called for no special actions are wanted: not even the capacity for continued attention is requisite. (Which is not to deny that from an intercultural point of view these demands can be seen as prodigious: the wonderful world of *Barry McKenzie* a comic strip is not apt to be available to those who haven't lived among the kangabloodyroos.) Popular art is popular because it is so readily available to all within the culture. But traditional works of art such as Leonardo's *Ginevra de 'Benci Mona Lisa* the madonna on the rocks Botticelli's Venus on the half shell are also popular and for much the same reason: from a western intracultural point of view an appreciation of these works calls for nothing special on the part of the viewer. The same is true of the appreciation of many carefully hand-crafted objects of many beautiful things in general.

When one turns to modern works demands on the person are apt to increase. Elliott Carter's *2nd Quartet* is a work of rare beauty but it is not instantly available to all. If one attempts in listening to the quartet to attend to recurring themes and variations as one would in listening to a work in standard sonata form then one is ready for Beethoven's *C minor Opus 18 No 4* but not for Carter: eighteen seconds of the opening *Allegro fantastico* should be enough to make that clear. Modern works of art often call for prolonged continuous close attention if one is to appreciate them. The same is true of a gator basking in the sun on a mud bank in a swamp. Anything viewed makes demands.

To suppose that anything that can be viewed is a fit object for aesthetic attention is not like supposing that anything one can put in one's mouth is a fit object to eat. It is more like supposing that anything that can be seen can be read. Because it can. It isn't true that one can't read just anything that one can see. Not everything has meaning but anything can be given meaning. One can read a blank piece of paper or a cloud or a sea anemone as some read palms and tea leaves and entrails. One can give meaning to stones but one can't make them edible. And one can see them as displays of solidity as expressive objects.

What's a fit subject to photograph? Anything that can be seen. Or is it not what the photographer photographs but what he makes of it? With his camera and darkroom and skills? What he does with art I can do with my (or maybe you too can with your) eyes. One can look at anything and within limits and depending on one's powers create an appropriate frame and environing conditions for what one sees.

I will describe what I call "antiaesthetic litter clearance". A nonaesthetic approach is a simple exercise in futility: the litter is offensive pick it up put it in trash cans sweep and tidy the area. Which owing to the unchanging propensities of the inhabitants will soon almost immediately be covered with litter again. The antiaesthetic approach is to alter one's view to see the original litter not as litter but as an object for aesthetic attention: a manifestation of a fundamental physical factor: entropy. One can look upon the disorder of litter as a form of order a beautiful randomness a precise display of imprecision. (And if you cannot look at litter in this way perhaps you can learn to do so by looking at Pollock Tobey and others.) Garbage strewn about is apt to be as delicately variegated in hue and value as the subtlest Monet. Discarded beer cans create striking cubistic patterns.

Consider a gator basking in the sun on a mud bank in a swamp. Is he a fit object for aesthetic attention? He is and that he is is readily confirmable. Go look and see if you doubt what I say. He is presently to be seen around Chokoloskee Island in the Everglades. What is in question is the American alligator (*Alligator mississipiensis*) not to be confused with a crocodile. Gators have shorter broader heads and more obtuse snouts. The fourth enlarged tooth of a gator's lower jaw fits into a pit formed for it in the upper jaw whereas a crocodile's fits into an external notch. It helps in viewing a gator to see it as a gator and not as a crocodile. But that requires knowing something about gators.

Seen from the side the gator appears to have a great healthy grin conveying a sense of well-being vitality. When Ginevra's portrait was painted by Leonardo she must have been sick for a long time. The pallor of her face conveys a "sense of melancholy".[1] The ossified scutes along his back forming the characteristic dermal armour constitute a powerful curving reticular pattern conveying simultaneously an impression of graceful fluidity and of remorseless solidity. Ginevra's face is "framed by cascading curls. These ringlets, infinitely varied in their shapes and movement, remind us of Leonardo's drawings of whirling eddies of water".[1] He has just come out of the water to bask in the sun. His sight is acute as is his power of hearing. But his eyes now have a lazy look being half-closed for he has upper and lower lids as well as a nictitating membrane. Ginevra too stares at us out of half-closed eyes. He is not strabismic. Her eyes are hazel. His seem green and remote despite the great grin.

Anything that can be viewed is a fit object for aesthetic attention. But not everything can be viewed just as not everything can be eaten. And in eating and in viewing the difficulties may be attributed either to the object or to the person. The former are obvious: stones can't be eaten and some gases subatomic particles and so forth can't be viewed because they can't be seen. But what cannot be eaten or cannot be viewed owing to the person is another matter. There are places where a rat foetus is considered a delicacy. The same is true of sheep's eye balls in aspic. In India warm monkey's brains are served up raw. Eskimos are reported to munch with delight on deer droppings (perhaps only in times of stress). Many in my society could not ingest these items: they would be stricken with nausea in the attempt. And there are hideous offensive nauseating objects that one cannot bear to view. Are such objects fit for aesthetic attention?

Yes why not? That I am psychologically incapable of attending aesthetically to a certain object tells you something about me nothing about the aesthetic qualities of the object. The same could be true of a work of art. Suppose Derain had done an heroic portrait of Hitler: I could not attend aesthetically to that work. Hitler was a repulsive nauseating object. That nausea is readily evoked by any lifelike image of the person. But my nausea would not be a criticism of Derain's art. Many of us cannot bear to look at blood particularly our own: that is not to deny that blood may be of a beautiful color and form beautiful patterns as it flows. If there were something that no one was psychologically capable of viewing even though the object was available for viewing then one might wonder whether such a thing was a fit object for aesthetic attention. But as far as I know there is no such thing and even if there were there's no need in theory anyway to countenance a morbid sensitivity that makes one psychologically incapable of viewing something in the world.

If anything that can be viewed is a fit object for aesthetic attention aren't some things more fit than others? No why think it? But granted that both a gator basking and Leonardo's *Ginevra* are fit objects for aesthetic attention isn't *Ginevra* more fit? No. In what way? It would make sense to compare the two only if there were some basis of comparison. But there isn't.

But isn't one painting better than another? In some ways and not in others. Rubens' paintings were superior to those of many of his contemporaries with respect to technique and pigmentation. Ingres' work displays finer draftsmanship than that of David. Vuillard's works have finer color than Manet's. But this isn't to say that Vuillard's works are more fit than those of Manet for aesthetic attention. If you are concerned to attend aesthetically to color then giving such attention to Vuillard's works will prove more worthwhile than giving such attention to Manet's works. But there are other things to attend to in viewing Manet's works. There are always other things to attend to.

For one can attend to anticolor: one can attend to precisely those aspects of hue value saturation of Manet's works which when standing on one's right foot adopting the stance of judge one judges to be inferior to Vuillard's. And one can without losing one's balance adopt a different stance standing on one's left foot one judges Manet's color superior to Vuillard's. And one can stand squarely on both feet and abandon the silliness of aesthetic judgements.

In looking at *Ginevra* one can attend to the display of craftsmanship and the beauty of form and shape: in looking at the gator basking one can attend to the beautiful grinning display of life. Anything that can be viewed can fill the bill of an object fit for aesthetic attention and none does it better than any other. Granted that 2 3 5 7 11 and so forth are primes: are some more so than others? A monk asked Ummon: "What is Buddha?" Ummon answered him: "Dried dung."

EPILOGUE

HOW I SEE PHILOSOPHY

Sometimes as questions. What does the word mean? What does the phrase mean? What does the sentence mean? What do you mean? Do you mean what your words mean? Do your words mean what you mean? Do you mean what your words mean but not what your phrases mean? Or do you mean what your words mean and what your phrases mean but not what your sentences mean? Or do you mean what your sentences mean but not what your words mean but what your phrases mean? Or do you mean what your sentences mean but not what your phrases mean but what your words mean? Or do you mean you mean what your sentences mean but not what your words and not what your phrases mean? Or do you mean that you mean what some of your words and all of your phrases but not your sentences mean? Sometimes as questions. Is there more than garbage in talk about necessity contingency possibility? Other than garbage in talk of intentions motives wants aims goals purposes? All kinds of questions any kind. What times is it? Where's George? Do you have anything to drink? Was the resemblance between Ike and Daddy Warbucks an accident? Because it's not the questions but how one thinks about it how one handles what one makes of it? Are there other minds? Yes. So pried at with simian fingers nothing interesting results for theory of some sort is what's wanted.

Sometimes as writings waves of words piled up spilling over leaking out of libraries dribbling out of corners of mouths. Russell Bradley Cook Wilson Bosanquet Prichard Lewis Langford Whitehead Bergson Wittgenstein Hegel Carnap Methodius Hutcheson Smith Shaftesbury Price Bentham Aquinas McTaggart Mctaggart Kant (thirty pages of whose pure reason ought to be enough to persuade even the reasonably well educated mentally deranged that it might be better not to turn to the past better not to dig up what is best left undug best not exhume the fetid remains of long ago postures) Dionysius the Pseudo-Areopagite Meister Eckhart Maimonides Gödel Peirce Little Orphan Annie Gregory Thaumaturgogo Herbrand Skolem the Green Hornet and the Shadow. Because it's not what one reads but what one makes of it? So we need and we gotta hab a deep hole foh to put in all de deep thinkers. Who we put in fust? It doan mattuh. Dump dem in! Der goes Kierkeebore. Bye Begel! So long Sartre! Dig dat Highdigger! Watchknow Will Durant? And dere's holy cow Playdough mould it ta any shape ya laks! C'mon gimme a paw ya pore ole clot ya doan b'long all de way down dere. Looka dat Weiss man! He fel on de haid o de Lip man drinkin Neerbeer. Christ man ya off ya rockuh? Pullin out Playdough an leavin an ole humean bean all Locked up! Yo is berkeleyin up de wrong.

Sometimes as resisting temptations. The temptation to commit history to mind. One doesn't read Plato Aristotle & Co without profit. The temptation to commit nomination: Ura analyst! Ura behaviourist! Ura a positivist! A third to categorize. "But that is an empirical question!" – this brought forth with a smile of innocence a pale expression wide open blue clear blue eyes the confidence with which these weighty words bind feet and hands roll the stone before and blocking and filling closing any gaping mouth.

Sometimes as an academic enterprise. For there are temptations everywhere crawling in every departmental chair. So as feet marching in three directions forever at odds with heads screwed on backwards mouths caught *in flagrante delicto* bright pink-red amber edged tongues rolling round and round and round and round and round and round.

Sometimes squinteyed distinctly presbyopic gazing at truths and as samples of savory sage metaphysical such we have. The first truth. Things get worse. Comment: this first truth is an exact statement of what is imprecisely expressed as the second law of thermodynamics.

Not all things all the time everywhere: there are apparent pockets of decreasing entropy but these are not closed systems etc. etc. etc. Things get worse. The second. For all x and for all y x is worse than y. Comment: it follows that x is worse than itself. But that's the way it is. The third. For all x the only good x is a dead x. Comment: is the relevant phrase a referring expression? The fourth. Which has a curious logical form. To the tutored eye it would seem to be a blend of declarative and interrogative forms: For all x and for all y who was dat x I seen yo wid? dat was no x dat was a y. But it is a true statement comment on reality appearance too.

Sometimes and mainly and mostly and for the most part and the main part and so principally and even chiefly as a janitor tending a conceptual zoo. Sweeping out categories combing concepts fighting fuddles cauterizing confusions pulling out monkey wrenches turning cages into fields finding fodder grinding raw beef into edible articled patties polishing tools and implements and instruments sorting sifting counting sand. But it's so hard to get anywhere when you have a neat an' orderly mind. The first thing you have to do of course the very first is to make sure that the slops stay in the bucket. That you don't empty them everywhere. You keep them in the bucket. You don't have to stir it. You just keep them in the bucket. That's the thing. You keep the slops in the bucket. Don't tip the bucket over! No! Don't stir the bucket. Just keep it in the corner. *Sum ergo cogito.*

NOTES

I

[1] *Topics in the Theory of Generative Grammar* (The Hague, Paris: Mouton, 1969), p. 11.

[2] John Berryman, *77 Dream Songs* (New York: Farrar, Straus and Giroux, 1965), p. 5.

[3] *Time*, Vol. 99, No. 8, Feb. 21 (1972), p. 8.

[4] See my *Understanding Understanding* (Ithaca: Cornell University Press, 1972), Chap. II for a discussion of "what is said".

[5] See my *Philosophic Turnings* (Ithaca: Cornell University Press, 1966), Chaps. XI–XIII for a discussion of related matters.

[6] D. B. Wyndham Lewis and Charles Lee (eds.), *The Stuffed Owl: An Anthology of Bad Verse* (New York: Capricorn Books, 1962), p. ix. This is the source of some of the verse cited in this essay.

[7] The wild Barbary sheep also called an aoudad.

[8] Peter Selz with texts by the artist, *The Work of Jean Dubuffet* (New York: The Museum of Modern Art, 1962), p. 97.

[9] *Op. cit.*, p. 103.

[10] John Skelton (1460–1529) wrote a funeral service for the sparrow Philip slain by a cat.

[11] All Souls College of Oxford University. Christopher Wren the room designer.

[12] From Andrew Marvell's poem of the nymph complaining of the death of her faun.

[13] Section XIX of the version edited by W. F. Stead.

[14] In T. S. Eliot's *Fragment of an Agon.*

[15] See T. Roethke's *The Meadow Mouse,* in his *The Far Field.*

[16] See M. Gell-Mann and E. P. Rosenbaum, 'Elementary Particles', *Scientific American* (July, 1957), p. 5 – my source for these comments.

[17] *Landscaped Tables Landscapes of the Mind Stones of Philosophy* (New York: Pierre Matisse Gallery, 1952), p. 3.

[18] See my *Understanding Understanding* (Ithaca: Cornell University Press, 1972), Chap. VIII.

II

[1] Darius Milhaud, *Notes without Music* (New York: Da Capo Press, 1970), pp. 101–102.

[2] E. T. Canby from the dustjacket for Nonesuch Records H–71122. Canby speaks of *The bull* as including "French music-hall tunes" but I have been unable to find any evidence for that claim. Nor do I know of any warrant for his remark that "'*Le Boeuf sur le toit*' was evidently the name of a Brazilian hostelry".

[3] See my *Understanding Understanding* (Ithaca: Cornell University Press, 1972), Chap. II.

[4] I am much indebted in music-theoretic matters to Marie Endres and Ann Woodward and Barbara Rowan.

III

[1] Constance Reid, *Hilbert* (New York, 1970), pp. 74–75.

[2] Hilbert was born on January 23, 1862.

[3] Norris and Ross McWhirter, *Guinness Sports Record Book*, 2nd Edition (New York, 1972), p. 112.

[4] *Ibid.*

[5] *Ibid.*

[6] Immanuel Kant, *Critique of Aesthetic Judgement*, Bk. *I 2*.

IV

[1] L. Wittgenstein, *Lectures and Conversations* (Berkeley and Los Angeles: University of California Press, 1967).

[2] Gyomay M. Kubose, *Zen Koans* (Chicago: Henry Regnery & Co., 1973), p. 75.

[3] *Science News*, Feb. 14 (1976), Vol. 109, No. 7, p. 101.

[4] *Theory of Self-Reproducing Automata* (Urbana and London: University of Illinois Press, 1966), p. 47.

VI

[1] E. O. Wilson, *On Human Nature* (Cambridge, Mass.: Harvard University Press, 1978), p. 16.

[2] M. Sahlins, *The Use and Abuse of Biology* (Ann Arbor: The University of Michigan Press, 1977), p. xi.

[3] Cited in Wilson, *op. cit.*, p. 21.

[4] *Cf.* P. Ziff, *Philosophic Turnings* (Ithaca: Cornell University Press, 1966), pp. 47–74.

[5] E. Mayr, *Evolution and the Diversity of Life* (Cambridge: Belknap Press, 1976), p. 518.

[6] *Cf.* Mayr, *op. cit.*, p. 144 ff.

[7] Mayr, *op. cit.*, p. 706.

[8] P. Handler (ed.), *Biology and the Future of Man* (New York: Oxford University Press, 1970), p. 466.

[9] *Op. cit.*, p. 389.

[10] Handler, p. 466.

[11] *Cf.* P. Ziff, 'Anything Viewed', chapter VIII of this volume.

[12] Mayr, p. 109.

[13] Mayr, p. 706.

[14] *Cf.* P. Ziff, 'Communication: Men and Other Animals' (forthcoming).

VII

[1] Norval Morris and Gordon Hawkins, *The Honest Politician's Guide to Crime Control* (Chicago and London: The University of Chicago Press, 1970), pp. 64–65.

[2] From some obsolete almanac: as usual things be gettin worse.

VIII

[1] Perry B. Cott, *Leonardo Da Vinci Ginevra de 'Benci* (Washington D.C.: National Gallery of Art, March, 1967).

BIBLIOGRAPHY (*of works referred to*)

Berryman, John. *77 Dream Songs.* New York: Farrar, Straus and Giroux, 1965.

Chomsky, Noam. *Topics in the Theory of Generative Grammar.* The Hague, Paris: Mouton, 1969.

Cott, Perry B. *Leonardo Da Vinci Ginevra de 'Benci.* Washington, D.C.: National Gallery of Art, 1967.

Dubuffet, Jean. *Landscaped Tables Landscapes of the Mind Stones of Philosophy.* New York: Pierre Matisse Gallery, 1952.

Eliot, T. S. *The Complete Poems and Plays.* New York: Harcourt, Brace and Company, 1952.

Gell-Mann, M. and Rosenbaum, E. P. "Elementary Particles," *Scientific American*, July, 1957.

Handler, P. (ed.) *Biology and the Future of Man.* New York: Oxford University Press, 1970.

Kant, Immanuel. *Critique of Aesthetic Judgement.* Oxford: The Clarendon Press, 1911.

Kubose, Gyomay M. *Zen Koans.* Chicago: Henry Regnery and Co., 1973.

Lewis, D. B. Wyndham and Lee, Charles (eds.) *The Stuffed Owl: An Anthology of Bad Verse.* New York: Capricorn Books, 1962.

Mayr, E. *Evolution and the Diversity of Life.* Cambridge: Belknap Press, 1976.

McWhirter, Norris and Ross. *Guinness Sports Record Book*, 2nd ed. New York: Sterling Pub. Co., 1972.

Milhaud, Darius. *Notes Without Music.* New York: Da Capo Press, 1970.

Morris, Norval and Hawkins, Gordon. *The Honest Politicians's Guide to Crime Control.* Chicago and London: The University of Chicago Press, 1970.

Reid, Constance. *Hilbert.* New York: Springer-Verlag, 1970.

Roethke, Theodore. *The Far Field.* Garden City, N.Y.: Doubleday and Company, Inc., 1964.

Sahlins, M. *The Use and Abuse of Biology.* Ann Arbor: The University of Michigan Press, 1977.

Selz, Peter. *The Work of Jean Dubuffet.* New York: The Museum of Modern Art, 1962.

Von Neumann, J. *Theory of Self-Reproducing Automata.* Urbana and London: University of Illinois Press, 1966.

Wilson, E. O. *On Human Nature.* Cambridge, Mass.: Harvard University Press, 1978.

Wittgenstein, L. *Lectures and Conversations.* Berkeley and Los Angeles: University of California Press, 1967.

Ziff, Paul. "Communication: Men and Other Animals", (forthcoming).

Ziff, Paul. *Philosophic Turnings.* Ithaca: Cornell University Press, 1966.

Ziff, Paul. *Understanding Understanding.* Ithaca: Cornell University Press, 1972.

INDEX

SYNTHESE LIBRARY

Studies in Epistemology, Logic, Methodology,
and Philosophy of Science

1. J. M. Bochénski, *A Precis of Mathematical Logic.* 1959.
2. P. L. Guiraud, *Problèmes et méthodes de la statistique linguistique.* 1960.
3. Hans Freudenthal (ed.), *The Concept and the Role of the Model in Mathematics and Natural and Social Sciences.* 1961.
4. Evert W. Beth, *Formal Methods. An Introduction to Symbolic Logic and the Study of Effective Operations in Arithmetic and Logic.* 1962.
5. B. H. Kazemier and D. Vuysje (eds.), *Logic and Language. Studies Dedicated to Professor Rudolf Carnap on the Occasion of His Seventieth Birthday.* 1962.
6. Marx W. Wartofsky (ed.), *Proceedings of the Boston Colloquium for the Philosophy of Science 1961-1962.* Boston Studies in the Philosophy of Science, Volume I. 1963.
7. A. A. Zinov'ev, *Philosophical Problems of Many-Valued Logic.* 1963.
8. Georges Gurvitch, *The Spectrum of Social Time.* 1964.
9. Paul Lorenzen, *Formal Logic.* 1965.
10. Robert S. Cohen and Marx W. Wartofsky (eds.), *In Honor of Philipp Frank.* Boston Studies in the Philosophy of Science, Volume II. 1965.
11. Evert W. Beth, *Mathematical Thought. An Introduction to the Philosophy of Mathematics.* 1965.
12. Evert W. Beth and Jean Piaget, *Mathematical Epistemology and Psychology.* 1966.
13. Guido Küng, *Ontology and the Logistic Analysis of Language. An Enquiry into the Contemporary Views on Universals.* 1967.
14. Robert S. Cohen and Marx W. Wartofsky (eds.), *Proceedings of the Boston Colloquium for the Philosophy of Science 1964-1966. In Memory of Norwood Russell Hanson.* Boston Studies in the Philosophy of Science, Volume III. 1967.
15. C. D. Broad, *Induction, Probability, and Causation. Selected Papers.* 1968.
16. Günther Patzig, *Aristotle's Theory of the Syllogism. A Logical-Philosophical Study of Book A of the Prior Analytics.* 1968.
17. Nicholas Rescher, *Topics in Philosophical Logic.* 1968.
18. Robert S. Cohen and Marx W. Wartofsky (eds.), *Proceedings of the Boston Colloquium for the Philosophy of Science 1966-1968.* Boston Studies in the Philosophy of Science, Volume IV. 1969.

19. Robert S. Cohen and Marx W. Wartofsky (eds.), *Proceedings of the Boston Colloquium for the Philosophy of Science 1966-1968.* Boston Studies in the Philosophy of Science, Volume V. 1969.
20. J. W. Davis, D. J. Hockney, and W. K. Wilson (eds.), *Philosophical Logic.* 1969.
21. D. Davidson and J. Hintikka (eds.), *Words and Objections. Essays on the Work of W. V. Quine.* 1969.
22. Patrick Suppes, *Studies in the Methodology and Foundations of Science. Selected Papers from 1911 to 1969.* 1969.
23. Jaakko Hintikka, *Models for Modalities. Selected Essays.* 1969.
24. Nicholas Rescher *et al.* (eds.), *Essays in Honor of Carl G. Hempel. A Tribute on the Occasion of His Sixty-Fifth Birthday.* 1969.
25. P. V. Tavanec (ed.), *Problems of the Logic of Scientific Knowledge.* 1969.
26. Marshall Swain (ed.), *Induction, Acceptance, and Rational Belief.* 1970.
27. Robert S. Cohen and Raymond J. Seeger (eds.), *Ernst Mach: Physicist and Philosopher.* Boston Studies in the Philosophy of Science, Volume VI. 1970.
28. Jaakko Hintikka and Patrick Suppes, *Information and Inference.* 1970.
29. Karel Lambert, *Philosophical Problems in Logic. Some Recent Developments.* 1970.
30. Rolf A. Eberle, *Nominalistic Systems.* 1970.
31. Paul Weingartner and Gerhard Zecha (eds.), *Induction, Physics, and Ethics.* 1970.
32. Evert W. Beth, *Aspects of Modern Logic.* 1970.
33. Risto Hilpinen (ed.), *Deontic Logic: Introductory and Systematic Readings.* 1971.
34. Jean-Louis Krivine, *Introduction to Axiomatic Set Theory.* 1971.
35. Joseph D. Sneed, *The Logical Sstructure of Mathematical Physics.* 1971.
36. Carl R. Kordig, *The Justification of Scientific Change.* 1971.
37. Milic Capek, *Bergson and Modern Physics.* Boston Studies in the Philosophy of Science, Volume VII. 1971.
38. Norwood Russell Hanson, *What I Do Not Believe, and Other Essays* (ed. by Stephen Toulmin and Harry Woolf). 1971.
39. Roger C. Buck and Robert S. Cohen (eds.), *PSA 1970. In Memory of Rudolf Carnap.* Boston Studies in the Philosophy of Science, Volume VIII. 1971.
40. Donald Davidson and Gilbert Harman (eds.), *Semantics of Natural Language.* 1972.
41. Yehoshua Bar-Hillel (ed.), *Pragmatics of Natural Languages.* 1971.
42. Sören Stenlund, *Combinators, λ-Terms and Proof Theory.* 1972.
43. Martin Strauss, *Modern Physics and Its Philosophy. Selected Papers in the Logic, History, and Philosophy of Science.* 1972.
44. Mario Bunge, *Method, Model and Matter.* 1973.
45. Mario Bunge, *Philosophy of Physics.* 1973.
46. A. A. Zinov'ev, *Foundations of the Logical Theory of Scientific Knowledge (Complex Logic).* (Revised and enlarged English edition with an appendix by G. A. Smirnov, E. A. Sidorenka, A. M. Fedina, and L. A. Bobrova.) Boston Studies in the Philosophy of Science, Volume IX. 1973.
47. Ladislav Tondl, *Scientific Procedures.* Boston Studies in the Philosophy of Science, Volume X. 1973.
48. Norwood Russell Hanson, *Constellations and Conjectures* (ed. by Willard C. Humphreys, Jr.). 1973.

49. K. J. J. Hintikka, J. M. E. Moravcsik, and P. Suppes (eds.), *Approaches to Natural Language.* 1973.
50. Mario Bunge (ed.), *Exact Philosophy – Problems, Tools, and Goals.* 1973.
51. Radu J. Bogdan and Ilkka Niiniluoto (eds.), *Logic, Language, and Probability.* 1973.
52. Glenn Pearce and Patrick Maynard (eds.), *Conceptual Change.* 1973.
53. Ilkka Niiniluoto and Raimo Tuomela, *Theoretical Concepts and Hypothetico-Inductive Inference.* 1973.
54. Roland Fraissé, *Course of Mathematical Logic* – Volume 1: *Relation and Logical Formula.* 1973.
55. Adolf Grünbaum, *Philosophical Problems of Space and Time.* (Second, enlarged edition.) Boston Studies in the Philosophy of Science, Volume XII. 1973.
56. Patrick Suppes (ed.), *Space, Time, and Geometry.* 1973.
57. Hans Kelsen, *Essays in Legal and Moral Philosophy* (selected and introduced by Ota Weinberger). 1973.
58. R. J. Seeger and Robert S. Cohen (eds.), *Philosophical Foundations of Science.* Boston Studies in the Philosophy of Science, Volume XI. 1974.
59. Robert S. Cohen and Marx W. Wartofsky (eds.), *Logical and Epistemological Studies in Contemporary Physics.* Boston Studies in the Philosophy of Science, Volume XIII. 1973.
60. Robert S. Cohen and Marx W. Wartofsky (eds.), *Methodological and Historical Essays in the Natural and Social Sciences. Proceedings of the Boston Colloquium for the Philosophy of Science 1969-1972.* Boston Studies in the Philosophy of Science, Volume XIV. 1974.
61. Robert S. Cohen, J. J. Stachel, and Marx W. Wartofsky (eds.), *For Dirk Struik. Scientific, Historical and Political Essays in Honor of Dirk J. Struik.* Boston Studies in the Philosophy of Science, Volume XV. 1974.
62. Kazimierz Ajdukiewicz, *Pragmatic Logic* (transl. from the Polish by Olgierd Wojtasiewicz). 1974.
63. Sören Stenlund (ed.), *Logical Theory and Semantic Analysis. Essays Dedicated to Stig Kanger on His Fiftieth Birthday.* 1974.
64. Kenneth F. Schaffner and Robert S. Cohen (eds.), *Proceedings of the 1972 Biennial Meeting, Philosophy of Science Association.* Boston Studies in the Philosophy of Science, Volume XX. 1974.
65. Henry E. Kyburg, Jr., *The Logical Foundations of Statistical Inference.* 1974.
66. Marjorie Grene, *The Understanding of Nature. Essays in the Philosophy of Biology.* Boston Studies in the Philosophy of Science, Volume XXIII. 1974.
67. Jan M. Broekman, *Structuralism: Moscow, Prague, Paris.* 1974.
68. Norman Geschwind, *Selected Papers on Language and the Brain.* Boston Studies in the Philosophy of Science, Volume XVI. 1974.
69. Roland Fraissé, *Course of Mathematical Logic* – Volume 2: *Model Theory.* 1974.
70. Andrzej Grzegorczyk, *An Outline of Mathematical Logic. Fundamental Results and Notions Explained with All Details.* 1974.
71. Franz von Kutschera, *Philosophy of Language.* 1975.
72. Juha Manninen and Raimo Tuomela (eds.), *Essays on Explanation and Understanding. Studies in the Foundations of Humanities and Social Sciences.* 1976.

73. Jaakko Hintikka (ed.), *Rudolf Carnap, Logical Empiricist. Materials and Perspectives.* 1975.
74. Milic Capek (ed.), *The Concepts of Space and Time. Their Structure and Their Development.* Boston Studies in the Philosophy of Science, Volume XXII. 1976.
75. Jaakko Hintikka and Unto Remes, *The Method of Analysis. Its Geometrical Origin and Its General Significance.* Boston Studies in the Philosophy of Science, Volume XXV. 1974.
76. John Emery Murdoch and Edith Dudley Sylla, *The Cultural Context of Medieval Learning.* Boston Studies in the Philosophy of Science, Volume XXVI. 1975.
77. Stefan Amsterdamski, *Between Experience and Metaphysics. Philosophical Problems of the Evolution of Science.* Boston Studies in the Philosophy of Science, Volume XXXV. 1975.
78. Patrick Suppes (ed.), *Logic and Probability in Quantum Mechanics.* 1976.
79. Hermann von Helmholtz: *Epistemological Writings. The Paul Hertz/Moritz Schlick Centenary Edition of 1921 with Notes and Commentary by the Editors.* (Newly translated by Malcolm F. Lowe. Edited, with an Introduction and Bibliography, by Robert S. Cohen and Yehuda Elkana.) Boston Studies in the Philosophy of Science, Volume XXXVII. 1977.
80. Joseph Agassi, *Science in Flux.* Boston Studies in the Philosophy of Science, Volume XXVIII. 1975.
81. Sandra G. Harding (ed.), *Can Theories Be Refuted? Essays on the Duhem-Quine Thesis.* 1976.
82. Stefan Nowak, *Methodology of Sociological Research. General Problems.* 1977.
83. Jean Piaget, Jean-Blaise Grize, Alina Szeminska, and Vinh Bang, *Epistemology and Psychology of Functions.* 1977.
84. Marjorie Grene and Everett Mendelsohn (eds.), *Topics in the Philosophy of Biology.* Boston Studies in the Philosophy of Science, Volume XXVII. 1976.
85. E. Fischbein, *The Intuitive Sources of Probabilistic Thinking in Children.* 1975.
86. Ernest W. Adams, *The Logic of Conditionals. An Application of Probability to Deductive Logic.* 1975.
87. Marian Przelecki and Ryszard Wójcicki (eds.), *Twenty-Five Years of Logical Methodology in Poland.* 1977.
88. J. Topolski, *The Methodology of History.* 1976.
89. A. Kasher (ed.), *Language in Focus: Foundations, Methods and Systems. Essays Dedicated to Yehoshua Bar-Hillel.* Boston Studies in the Philosophy of Science, Volume XLIII. 1976.
90. Jaakko Hintikka, *The Intentions of Intentionality and Other New Models for Modalities.* 1975.
91. Wolfgang Stegmüller, *Collected Papers on Epistemology, Philosophy of Science and History of Philosophy.* 2 Volumes. 1977.
92. Dov M. Gabbay, *Investigations in Modal and Tense Logics with Applications to Problems in Philosophy and Linguistics.* 1976.
93. Radu J. Bogdan, *Local Induction.* 1976.
94. Stefan Nowak, *Understanding and Prediction. Essays in the Methodology of Social and Behavioral Theories.* 1976.
95. Peter Mittelstaedt, *Philosophical Problems of Modern Physics.* Boston Studies in the Philosophy of Science, Volume XVIII. 1976.

96. Gerald Holton and William Blanpied (eds.), *Science and Its Public: The Changing Relationship*. Boston Studies in the Philosophy of Science, Volume XXXIII. 1976.
97. Myles Brand and Douglas Walton (eds.), *Action Theory*. 1976.
98. Paul Gochet, *Outline of a Nominalist Theory of Proposition. An Essay in the Theory of Meaning*. 1980.
99. R. S. Cohen, P. K. Feyerabend, and M. W. Wartofsky (eds.), *Essays in Memory of Imre Lakatos*. Boston Studies in the Philosophy of Science, Volume XXXIX. 1976.
100. R. S. Cohen and J. J. Stachel (eds.), *Selected Papers of Léon Rosenfeld*. Boston Studies in the Philosophy of Science, Volume XXI. 1978.
101. R. S. Cohen, C. A. Hooker, A. C. Michalos, and J. W. van Evra (eds.), *PSA 1974: Proceedings of the 1974 Biennial Meeting of the Philosophy of Science Association*. Boston Studies in the Philosophy of Science, Volume XXXII. 1976.
102. Yehuda Fried and Joseph Agassi, *Paranoia: A Study in Diagnosis*. Boston Studies in the Philosophy of Science, Volume L. 1976.
103. Marian Przelecki, Klemens Szaniawski, and Ryszard Wójcicki (eds.), *Formal Methods in the Methodology of Empirical Sciences*. 1976.
104. John M. Vickers, *Belief and Probability*. 1976.
105. Kurt H. Wolff, *Surrender and Catch: Experience and Inquiry Today*. Boston Studies in the Philosophy of Science, Volume LI. 1976.
106. Karel Kosík, *Dialectics of the Concrete*. Boston Studies in the Philosophy of Science, Volume LII. 1976.
107. Nelson Goodman, *The Structure of Appearance*. (Third edition.) Boston Studies in the Philosophy of Science, Volume LIII. 1977.
108. Jerzy Giedymin (ed.), *Kazimierz Ajdukiewicz: The Scientific World-Perspective and Other Essays, 1931-1963*. 1978.
109. Robert L. Causey, *Unity of Science*. 1977.
110. Richard E. Grandy, *Advanced Logic for Applications*. 1977.
111. Robert P. McArthur, *Tense Logic*. 1976.
112. Lars Lindahl, *Position and Change. A Study in Law and Logic*. 1977.
113. Raimo Tuomela, *Dispositions*. 1978.
114 Herbert A. Simon, *Models of Discovery and Other Topics in the Methods of Science*. Boston Studies in the Philosophy of Science, Volume LIV. 1977.
115. Roger D. Rosenkrantz, *Inference, Method and Decision*. 1977.
116. Raimo Tuomela, *Human Action and Its Explanation. A Study on the Philosophical Foundations of Psychology*. 1977.
117. Morris Lazerowitz, *The Language of Philosophy. Freud and Wittgenstein*. Boston Studies in the Philosophy of Science, Volume LV. 1977.
118. Stanislaw Leśniewski, *Collected Works* (ed. by S. J. Surma, J. T. J. Srzednicki, and D. I. Barnett, with an annotated bibliography by V. Frederick Rickey). 1982. (Forthcoming.)
119. Jerzy Pelc, *Semiotics in Poland, 1894-1969*. 1978.
120. Ingmar Pörn, *Action Theory and Social Science. Some Formal Models*. 1977.
121. Joseph Margolis, *Persons and Minds. The Prospects of Nonreductive Materialism*. Boston Studies in the Philosophy of Science, Volume LVII. 1977.
122. Jaakko Hintikka, Ilkka Niiniluoto, and Esa Saarinen (eds.), *Essays on Mathematical and Philosophical Logic*. 1978.
123. Theo A. F. Kuipers, *Studies in Inductive Probability and Rational Expectation*. 1978.

124. Esa Saarinen, Risto Hilpinen, Ilkka Niiniluoto, and Merrill Provence Hintikka (eds.), *Essays in Honour of Jaakko Hintikka on the Occasion of His Fiftieth Birthday.* 1978.
125 Gerard Radnitzky and Gunnar Andersson (eds.), *Progress and Rationality in Science.* Boston Studies in the Philosophy of Science, Volume LVIII. 1978.
126. Peter Mittelstaedt, *Quantum Logic.* 1978.
127. Kenneth A. Bowen, *Model Theory for Modal Logic. Kripke Models for Modal Predicate Calculi.* 1978.
128. Howard Alexander Bursen, *Dismantling the Memory Machine. A Philosophical Investigation of Machine Theories of Memory.* 1978.
129. Marx W. Wartofsky, *Models: Representation and the Scientific Understanding.* Boston Studies in the Philosophy of Science, Volume XLVIII. 1979.
130. Don Ihde, *Technics and Praxis. A Philosophy of Technology.* Boston Studies in the Philosophy of Science, Volume XXIV. 1978.
131. Jerzy J. Wiatr (ed.), *Polish Essays in the Methodology of the Social Sciences.* Boston Studies in the Philosophy of Science, Volume XXIX. 1979.
132. Wesley C. Salmon (ed.), *Hans Reichenbach: Logical Empiricist.* 1979.
133. Peter Bieri, Rolf-P. Horstmann, and Lorenz Krüger (eds.), *Transcendental Arguments in Science. Essays in Epistemology.* 1979.
134. Mihailo Marković and Gajo Petrović (eds.), *Praxis. Yugoslav Essays in the Philosophy and Methodology of the Social Sciences.* Boston Studies in the Philosophy of Science, Volume XXXVI. 1979.
135. Ryszard Wójcicki, *Topics in the Formal Methodology of Empirical Sciences.* 1979.
136. Gerard Radnitzky and Gunnar Andersson (eds.), *The Structure and Development of Science.* Boston Studies in the Philosophy of Science, Volume LIX. 1979.
137. Judson Chambers Webb. *Mechanism, Mentalism, and Metamathematics. An Essay on Finitism.* 1980.
138. D. F. Gustafson and B. L. Tapscott (eds.), *Body, Mind, and Method. Essays in Honor of Virgil C. Aldrich.* 1979.
139. Leszek Nowak, *The Structure of Idealization. Towards a Systematic Interpretation of the Marxian Idea of Science.* 1979.
140. Chaim Perelman, *The New Rhetoric and the Humanities. Essays on Rhetoric and Its Applications.* 1979.
141. Wlodzimierz Rabinowicz, *Universalizability. A Study in Morals and Metaphysics.* 1979.
142. Chaim Perelman, *Justice, Law, and Argument. Essays on Moral and Legal Reasoning.* 1980.
143. Stig Kanger and Sven Öhman (eds.), *Philosophy and Grammar. Papers on the Occasion of the Quincentennial of Uppsala University.* 1981.
144. Tadeusz Pawlowski, *Concept Formation in the Humanities and the Social Sciences.* 1980.
145. Jaakko Hintikka, David Gruender, and Evandro Agazzi (eds.), *Theory Change, Ancient Axiomatics, and Galileo's Methodology. Proceedings of the 1978 Pisa Conference on the History and Philosophy of Science, Volume I.* 1981.
146. Jaakko Hintikka, David Gruender, and Evandro Agazzi (eds.), *Probabilistic Thinking, Thermodynamics, and the Interaction of the History and Philosophy of*

Science. Proceedings of the 1978 Pisa Conference on the History and Philosophy of Science, Volume II. 1981.

147. Uwe Mönnich (ed.), *Aspects of Philosophical Logic. Some Logical Forays into Central Notions of Linguistics and Philosophy.* 1981.
148. Dov M. Gabbay, *Semantical Investigations in Heyting's Intuitionistic Logic.* 1981.
149. Evandro Agazzi (ed.), *Modern Logic – A Survey. Historical, Philosophical, and Mathematical Aspects of Modern Logic and its Applications.* 1981.
150. A. F. Parker-Rhodes, *The Theory of Indistinguishables. A Search for Explanatory Principles below the Level of Physics.* 1981.
151. J. C. Pitt, *Pictures, Images, and Conceptual Change. An Analysis of Wilfrid Sellars' Philosophy of Science.* 1981.
152. R. Hilpinen (ed.), *New Studies in Deontic Logic. Norms, Actions, and the Foundations of Ethics.* 1981.
153. C. Dilworth, *Scientific Progress. A Study Concerning the Nature of the Relation Between Successive Scientific Theories.* 1981.
154. D. W. Smith and R. McIntyre, *Husserl and Intentionality. A Study of Mind, Meaning, and Language.* 1982.
155. R. J. Nelson, *The Logic of Mind.* 1982.
156. J. F. A. K. van Benthem, *The Logic of Time. A Model-Theoretic Investigation into the Varieties of Temporal Ontology, and Temporal Discourse.* 1982.
157. R. Swinburne (ed.), *Space, Time and Causality.* 1982.
158. R. D. Rozenkrantz, *E. T. Jaynes: Papers on Probability, Statistics and Statistical Physics.* 1983.
159. T. Chapman, *Time: A Philosophical Analysis.* 1982.
160. E. N. Zalta, *Abstract Objects. An Introduction to Axiomatic Metaphysics.* 1983.
161. S. Harding and M. B. Hintikka (eds.), *Discovering Reality. Feminist Perspectives on Epistemology, Metaphysics, Methodology, and Philosophy of Science.* 1983.
162. M. A. Stewart (ed.), *Law, Morality and Rights.* 1983.
163. D. Mayr and G. Süssmann (eds.), *Space, Time, and Mechanics. Basic Structure of a Physical Theory.* 1983.
164. D. Gabbay and F. Guenthner (eds.), *Handbook of Philosophical Logic.* Vol. I. 1983.
165. D. Gabbay and F. Guenthner (eds.), *Handbook of Philosophical Logic.* Vol. II, 1984.
166. D. Gabbay and F. Guenthner (eds.), *Handbook of Philosophical Logic.* Vol. III, forthcoming.
167. D. Gabbay and F. Guenthner (eds.), *Handbook of Philosophical Logic.* Vol.IV, forthcoming.
168. Andrew, J. I. Jones, *Communication and Meaning.* 1983.
169. Melvin Fitting, *Proof Methods for Modal and Intuitionistic Logics.* 1983.
170. Joseph Margolis, *Culture and Cultural Entities.* 1984.
171. Raimo Tuomela, *A Theory of Social Action.* 1984.
172. Jorge J. E. Gracia, Eduardo Rabossi, Enrique Villanueva, and Marcelo Dascal (eds.), *Philosophical Analysis in Latin America.* 1984.
173. Paul Ziff, *Epistemic Analysis. A Coherence Theory of Knowledge.* 1984.
174. Paul Ziff, *Antiaesthetics. An Appreciation of the Cow with the Subtile Nose.* 1984.

175. Wolfgang Balzer, David A. Pearce, and Heinz-Jürgen Schmidt (eds.), *Reduction in Science. Structure, Examples, Philosophical Problems.* 1984.
176. Aleksander Peczenik, Lars Lindahl, and Bert van Roermund (eds.), *Theory of Legal Science. Proceedings of the Conference on Legal Theory and Philosophy of Science, Lund, Sweden, December 11–14, 1983*, forthcoming.
177. Ilkka Niniluoto, *Is Science Progressive?* 1984.
178. Bimal Matilal and Jaysankar Lal Shaw (eds.), *Exploratory Essays in Current Theories and Classical Indian Theories of Meaning and Reference*, forthcoming.
179. Peter Kroes, *Time: Its Structure and Role in Physical Theories,* forthcoming.